Psychology Demystified

Demystified Series

Accounting Demystified
Advanced Statistics Demystified
Algebra Demystified
Alternative Energy Demystified
Anatomy Demystified
asp.net 2.0 Demystified
Astronomy Demystified
Audio Demystified
Biology Demystified
Biotechnology Demystified
Business Calculus Demystified
Business Math Demystified
Business Statistics Demystified
C++ Demystified
Calculus Demystified
Chemistry Demystified
College Algebra Demystified
Corporate Finance Demystified
Data Structures Demystified
Databases Demystified
Differential Equations Demystified
Digital Electronics Demystified
Earth Science Demystified
Electricity Demystified
Electronics Demystified
Environmental Science Demystified
Everyday Math Demystified
Forensics Demystified
Genetics Demystified
Geometry Demystified
Home Networking Demystified
Investing Demystified
Java Demystified
JavaScript Demystified
Linear Algebra Demystified
Macroeconomics Demystified
Management Accounting Demystified

Math Proofs Demystified
Math Word Problems Demystified
Medical Billing and Coding Demystified
Medical Terminology Demystified
Meteorology Demystified
Microbiology Demystified
Microeconomics Demystified
Nanotechnology Demystified
Nurse Management Demystified
OOP Demystified
Options Demystified
Organic Chemistry Demystified
Personal Computing Demystified
Pharmacology Demystified
Physics Demystified
Physiology Demystified
Pre-Algebra Demystified
Precalculus Demystified
Probability Demystified
Project Management Demystified
Psychology Demystified
Quality Management Demystified
Quantum Mechanics Demystified
Relativity Demystified
Robotics Demystified
Signals and Systems Demystified
Six Sigma Demystified
sql Demystified
Statics and Dynamics Demystified
Statistics Demystified
Technical Math Demystified
Trigonometry Demystified
uml Demystified
Visual Basic 2005 Demystified
Visual C# 2005 Demystified
xml Demystified

Psychology Demystified

ANNA ROMERO, PH.D.
STEVEN M. KEMP, PH.D.

New York   Chicago San Francisco Lisbon London
Madrid   Mexico City Milan New Delhi San Juan
Seoul Singapore Sydney Toronto

The McGraw·Hill Companies

McGraw-Hill books are available at special quantity discounts to use as premiums and sales promotions, or for use in corporate training programs. For more information, please write to the Director of Special Sales, Professional Publishing, McGraw-Hill, Two Penn Plaza, New York, NY 10121-2298. Or contact your local bookstore.

Psychology Demystified

1234567890 DOC DOC 019876

ISBN-13:   978-0-07-146030-9
ISBN-10:   0-07-146030-6

Sponsoring Editor
Judy Bass

Editorial Supervisor
Janet Walden

Project Manager
Samik Roy Chowdhury (Sam)

Copy Editor
LeeAnn Pickrell

Proofreader
Julie Searls

Indexer
WordCo Indexing Services

Production Supervisor
Jean Bodeaux

Composition
International Typesetting and Composition

Illustration
Adam Brill

Art Director, Cover
Margaret Webster-Shapiro

Cover Illustration
Lance Lekander

To Sam Fillenbaum, who revealed the mysteries of psychology to both of us.

ABOUT THE AUTHORS

Anna Romero, Ph.D., has taught introductory psychology, among other courses, at several universities. As General Psychology Fellow at the University of North Carolina at Chapel Hill, she managed the department-wide coordination of the introductory psychology curriculum, and the training of junior graduate teaching assistants. Dr. Romero recently completed a four-year postdoctoral research position in the Psychology departments at New York University and Princeton. She is currently a researcher with Harris Interactive.

Steven M. Kemp, Ph.D., is a Research Assistant Professor at the University of North Carolina at Chapel Hill, where he has taught classes in introductory psychology as well as courses in experimental psychology and psychological methods. Professor Kemp is a member of the American Psychological Association, the American Psychological Society, and the Association for Behavior Analysis. He is co-author of *Business Statistics Demystified* from McGraw-Hill (2004).

CONTENTS AT A GLANCE

CONTENTS

Contents

Contents

PREFACE

Psychology Demystified is designed both for students in introductory psychology classes as well as for anyone who wants to find out about today's psychology in the lab and in the clinic. We have matched the chapter topics of the current major (and minor) textbooks to make things easy for students working with their assigned text, as well as to ensure complete coverage.

Everyone has his or her own ideas about what makes people tick. It's not always easy to understand how psychologists study people from their point of view. There are many textbooks on introductory psychology and many books, such as *Psychology Demystified*, that are designed to cover things more informally. Almost all of these books work hard to make psychology accessible. If you find these books tough going, it might be because, in order to keep things simple, many of these books provide only one perspective—the view of psychology shared by most of the authors. We have our own views, of course, but with *Psychology Demystified*, we have worked hard to give you a broader view than just our own. For each topic, we show how different psychologists think about the topic in different ways. That makes it easier to see how your own ideas about people fit in with how psychologists study people scientifically.

Psychology is a young science. Many of our theories and even our experiments are still controversial. You might find it easier to understand some of it from one side of the controversy than from the other. That's why we tell you both sides of the story. When there are alternatives to the mainstream view, we let you know about it.

Psychologists are scientists. They take a properly scientific, skeptical attitude toward their subject matter. However, many of us aren't so skeptical when it comes to the science of psychology itself. Introductory texts, in particular, tend to have a gung-ho, pep-rally attitude about just how good a job psychology is doing, both in the laboratory and in the clinic. But all of that success is because of how psychologists think about people and what questions they think are important. Until you can see psychology's point of view, the record of our success might not make much sense. Here, we don't just tell you the answers psychologists have found, but why they think those particular questions are important in the first place.

Psychology Demystified is for you if

- *You are in an introductory psychology class, and you find it challenging.* Whether you just can't seem to think like a psychologist, or you're not sure what the problem is, the answer is here. You will see how psychologists think, and you will be able to relate it to your own understanding of people.

- *You are in an introductory psychology class, and you want to excel.* Psychology is a very diverse field. Keeping track of the terminology and the long lists of experimental results is difficult. We show you the big picture and how the details fit in, which makes it a lot easier to remember.

- *You want to find out about psychology on your own.* We tell the story of how psychology got where it is, which is a lot more fun than reading a long list of what psychologists have discovered in the laboratory.

We wrote this book so that you can get a broad understanding of how psychologists study people. When you have finished reading this book, you will find that you can

- View and discuss behavior from a scientific perspective

- Understand and evaluate media reports of psychological studies

- See how psychologists divide up the various fields of psychology

- Understand the advantages and limitations of the scientific method for studying people

- Relate what psychologists have found out about human behavior to your own understanding of people

- Understand the wide panorama of different perspectives on the many different aspects of human behavior studied by psychologists

- Understand more about the relationship between the mind and the brain

- Know about the history of psychology

- Avoid the most common misunderstandings about psychology and psychologists

How to Get the Most Out of This Book

No matter what parts of psychology you are most interested in, you should probably start at the beginning and work your way through Chapter 1. This will give you a better understanding of how psychologists view things. After that, you can skip around. If you are taking a class, you will find that our chapter topics match up easily with the chapters in your textbook, so you can read them in the order assigned in your class. In fact, you might read the overview to each chapter in this book before you look at the corresponding chapter in your textbook, so you can avoid getting mystified in the first place!

If you are reading *Psychology Demystified* on its own, you might want to pick and choose chapters on topics that interest you after you have gotten through Chapter 1. Or you might want to check out the story of psychology in the order we tell it. It's a bit different than the way other recent authors have done it.

Sidebars and Tests for Easy Learning

In *Psychology Demystified*, we want to make it easy for you to learn and to find what you need to know. So we've created several different types of sidebars that will introduce key ideas:

- **Tips on Terms**   Definitions and explanations of crucial terminology.
- **Critical Cautions**   Common mistakes to avoid in order to get things right.
- **Study Review**   Key points for exam preparation.
- **Handy Hints**   Practical advice on how to understand a topic.
- **Fun Facts**   A little bit on the lighter side.
- **Variant Views**   Other ways that psychologists have thought about things that don't always make it into the textbooks.
- **Big Background**   A step back to look at the history of how things got where they are.
- **Key Points**   Highlighting and explaining important concepts.
- **Reading Rules**   Guidance on how to approach a part of the text.
- **Case Study**   A specific historical example illustrating a concept.
- **Topical Talk**   A question for discussion and debate.

- **Bio Bites** The authors' experience. We've seen a few things.
- **Quick Quotes** Bits of wisdom from folks much smarter than we are.

As with all books in the *Demystified* series, we have provided quizzes at the end of each chapter, tests at the end of each part of the book, and a final exam. The test questions (and the answers at the back) will not only help you learn the facts, but also be good practice for your tests if you are taking a class.

ACKNOWLEDGMENTS

Our first thanks go to Adam Brill, our wonderful illustrator, who's impeccably professional work made for all the illustrations in *Psychology Demystified*, except for the diagram on the theories of emotion. Then there's the second author's brother, Sid Kemp, who came up with the idea of *Psychology Demystified* and convinced us we should be the ones to write it. Sid also provided us with the basic organizational materials we used in authoring the book and in keeping track of all the files as they bounced back and forth over the Internet from Oklahoma to North Carolina. Sid has written a lot more books than we have and was generous with his advice. (He also did a bang-up job on that emotions diagram.)

Dave Eckerman, Dick King, Paul Shinkman, James W. Kalat, Jeannie Koo-Loeb, and Ed Morrill all helped us obtain various source materials. Sam Fillenbaum and Joseph Lowman helped us with recommended readings. Carlos Diuk made sure the mathematical curves in our figures and diagrams were exact and beautiful. Eva Scardina helped assemble the answer key for the quizzes, tests, and the exam. Kelila Feigenbaum copyedited our initial manuscript, prepared the references, and took all the quizzes, tests, and the exam in order to cross-check the answer key. Jessi Perlson helped put together the glossary. And, of course, special thanks to Judy Bass, the *Demystified* series editor at McGraw-Hill, whose support and near-infinite patience kept us going.

The first author would like to thank her family for all of their support, emotional and otherwise. Bret Wing, Ph.D. also deserves thanks for first, warning me that writing a book is both an exhilarating and difficult process, and second, for encouraging me every step of the way. The Cincinnati crew of Harris Interactive deserve a big thanks for giving me a new place to call "work" and for being so patient with my publishing deadlines. Finally, Michael Cox and Charles Hunt both deserve thanks for reminding me to put down the manuscript once in a while and have a good meal.

The second author would like to thank Joy Preslar, Melanie Martin, and Jessi Perlson, who helped me get my personal library (and my life) in order so I could *find* all those lovely source materials. Dave Eckerman and Peter Ornstein, both of the Psychology Department at the University of North Carolina at Chapel Hill, have supported my affiliation with that institution, whose extensive research resources were invaluable in the preparation of the manuscript of the book.

Finally, both of us want to thank all of our teachers. There is an awful lot of psychology in this book, and we had an awful lot of wonderful teachers who taught it to us so we could demystify it for you.

INTRODUCTION

What Is Psychology?

Psychology is often defined as the scientific study of behavior (or of behavior and the mind). This definition has the benefit of distinguishing psychology from history, literature, sculpture, religion, and other nonscientific ways of looking at behavior. Other scientific studies of people, such as economics, anthropology, ethology, and sociology, are distinguished from psychology because they have a narrower focus.

The use of the words, *mind* and *behavior* are something of a political compromise. Some psychologists think of *mind* as a prescientific term for the organization of behavior. Some psychologists think of *behavior* as the mere outward manifestations of the true subject of psychological study. Terminology aside, there is little debate as to what psychologists actually study. The actions people take are uniquely organized to allow us to cope with the world in ways that vegetables, minerals, and other animals cannot. Psychologists study those various activities and the uniquely human abilities that make them possible.

Tips on Terms

Psychologists actually define *scientific* more narrowly than other scientists do. Many techniques used in other sciences are less acceptable in modern psychology. The field studies common to anthropology are rarely found in psychology. Observational techniques, such as those used in astronomy and archeology, also have limited use. Theories, at least in the sense that they exist in sciences such as physics, do not really exist in psychology. (Throughout this book, we will see many discussions of psychological "theories," but these are much narrower in focus than the grand theories found in contemporary physics, and they play a very different role.)

Psychology as a Discipline

People have always wondered about human behavior and that includes people in the academy. Plato worked out various ideas that inform our modern conception of the mind. The first applications of the scientific method were to the physical, rather than the psychological world. It was August Comte (1798–1857) who first proposed that the scientific method could be applied to the study of people.

HOW DID PSYCHOLOGY BECOME A SEPARATE DISCIPLINE?

For decades thereafter, the people who attempted to apply the scientific method to the study of people were found in philosophy departments at universities. There were also biologists and physicians studying other animals with an eye toward understanding people. It was not until the 1870s in Germany (and the 1880s in the United States) that psychology departments and psychology laboratories began to be established. By the 1920s, most large universities had separate psychology departments, with similar areas of study to what they have now.

FROM THE UNIVERSITY TO THE CLINIC

In the United States, the federal government began taking an interest in the practical applications of psychology as early as World War I. After the success of the Manhattan Project in World War II, the Feds decided that the best way to wage the Cold War was to invest in science. The long association between psychology and the federal government, especially the military, paid off. A number of government offices dedicated to funding scientific research allocated substantial monies to the advancement of psychology.

Up until this time, the psychologists' role in treating people with mental problems had been mostly subsidiary to the therapies usually supplied by psychiatrists. (Psychiatrists are physicians who specialize in treating mental ailments. They may or may not have psychological training.) Psychologists who worked with psychiatrists often specialized in giving various sorts of tests to patients. These tests assessed the patients' intelligence or level of functioning or helped with diagnosis. Few psychologists offered psychotherapy on their own.

The huge influx of funds to psychology and psychologists after World War II came from a federal government not only interested in practical solutions to practical problems generally, but whose long experience with psychology had been in practical areas. Much of this money was directed toward applied research and training of nonacademic psychologists. As a result, there was an enormous expansion of

the number of psychologists working in the real world instead of just the university setting. The vast majority of these psychologists were trained to provide psychotherapy to patients in a clinical setting.

Within a few decades, the American Psychological Association, the main organization for psychologists, went from having a small minority of clinicians to being 80 percent clinicians. It also expanded enormously, reaching approximately 160,000 members by the end of the twentieth century. Gradually, during the same time period, the number and variety of other applied psychological practitioners also expanded greatly.

Philosophical Concerns

The application of the scientific method to psychology has always been a challenge. The scientific method is designed to reveal our mistaken assumptions and replace our myths with facts. People are always reluctant to give up their myths. And psychology is the study of *us*. People are especially reluctant to give up their myths about themselves.

ISSUES AND DISPUTES

It was Darwin, with his view that people were just another type of animal, who made the scientific study of people possible. To this day, many people are still uncomfortable with Darwin's notion that people are animals. Psychology not only follows in the Darwinian tradition, but its main focus is on those aspects of human behavior that we do not share with other animals. Explaining what is uniquely human in scientific terms can be even more threatening to our pride than Darwin's explaining what we have in common with other animals in scientific terms.

One simple way to deal with this is to reject the notion of psychology entirely. Another way is to try to structure psychology so that it preserves our older ideas about ourselves. Many of the philosophical issues surrounding psychology can be linked to this second way of protecting our image of ourselves.

APPROACHES

Prior to the advent of psychology, philosophers had a long history of developing an extensive picture of the mind. The first psychologists, called *structuralists*, took their primary responsibility to be the answering of age-old questions from the philosophy of mind that had not succumbed to philosophical inquiry but might fall to scientific study. This was precisely how physics got started, although, in the end, physics had to reject the questions that philosophers had posed, rather than answer them.

Structuralism did not last long. It relied upon the method of *introspection* (basically, trusting the human subjects to be able to say exactly what was going on in their minds). Beginning in the early part of the twentieth century, structuralism was replaced with *behaviorism*, which rejected the very notion of mind, along with many other traditional philosophical concerns. (Rejecting old philosophies was very chic in the twentieth century. This antiphilosophy philosophy was called *logical positivism*.)

Behaviorism made considerable progress, but collapsed rapidly in the 1960s, in part because it was cumbersome in addressing the notion of the mind, in part because it took a cautious approach to addressing the most pressing practical concerns of society, but also in part because its antiphilosophy philosophy collapsed. There are still some behaviorists about, but they are almost exclusively followers of B. F. Skinner, who rejected logical positivism in 1945.

Currently, the most popular approach to psychology is called *cognitivism*. Cognitivism retains the methodology of behaviorism, along with a commitment to the idea that nothing about the mind is beyond the physical. The mind, according to cognitivists, is real, but it is just a property of the human brain, not a separate non-physical spirit. The philosophical core to cognitivism is the notion that very old, very traditional notions of the mind dating back to Plato can be reconciled to biological facts about the brain, rendering the mind a valid subject of scientific study.

The keys to this proposed reconciliation are two ideas taken from computers, *process* and *representation*. Computer processes are controlled by highly structured lists of rules, called programs. This fact demonstrates that a structure can determine a process, even the most complex process. The brain has a very complex structure, and cognitivists understand the mind as a complex process determined by that structure. It is the notion of representation, however, that ensures that mental processes have the traditional character discussed by philosophers for so many centuries.

Representations are symbols inside a process that stand in for real objects and events outside of the process. The central philosophical doctrine of cognitivism is that, when we think about something, symbols in our brain are representing the things we are thinking about. As we might imagine, this doctrine (called *representationalism*) is extremely hard to demonstrate in the laboratory. So long as philosophers continue to advocate the underlying philosophy, cognitivism will continue.

Psychology's Subfields

Psychology is a very diverse field. In no small part, this is because human behavior is highly varied. To some degree, however, it is also because the basic principles underlying human behavior (if there are any) are still controversial. Unlike physics or chemistry or even biology or economics, there is, as yet, no core set of principles that guide all of the different parts of psychology. Psychology is joined together by a common interest in behavior and the specialized scientific method for understanding it.

For simplicity, we divide the areas of psychology into basic and applied. Basic areas are concerned with inquiry into what, how, and why the human mind works as it does. Applied areas are concerned with making use of the knowledge gained in the basic areas to better the human situation.

BASIC

Here are some different areas of basic psychology:

- *Social psychology* is the study of the effect of social interaction on behavior.
- *Developmental psychology* is the study of the changes in behavior due to age.
- *Psychological testing* is the study of the measurement of human abilities, including intelligence.
- *Personality psychology* is the study of individual differences in behavior not related to age or ability.
- The *psychology of consciousness* is the study of our experience of ourselves.
- The *psychology of emotion* is the study of the feelings that color our experience.
- The *psychology of motivation* is the study of why we do what we do.
- The *psychology of learning* is the study of how experience changes us to be more capable in the world.
- *Psycholinguistics* is the study of how people communicate using language.
- *Cognitive psychology* is the study of thinking. It is closely related to the study of the senses and of perceiving the world.
- *Psychophysics* is the study of how the physical world creates the sensory world.

- *Neuroscience* (formerly biological psychology) attempts to relate behavior to the brain function that causes it.
- *Psychometrics/quantitative psychology* is the study of research methods in psychology. These are the tool-builders for the rest of psychology.

APPLIED

Like any science, the ultimate demonstration that we got things right is when we are able to use the knowledge derived from scientific psychology to change the world. Some psychologists research how to change the world; others actually go out and do it.

Research

Here are some different areas of applied research in psychology:

- *Industrial/organizational psychology* is the study of psychological aspects of business.
- *Educational psychology* is the study of the psychology of teaching.
- *Abnormal psychology* is the study of mental disorders.
- *Psychotherapy* is the study of ways of treating people with mental disorders.
- *Psychopharmacology* is the study of the effects of drugs on behavior.
- *Health psychology* is the study of the interaction between behavior and health.

Practice

Here are some different areas of psychological practice:

- *School psychologists* work in schools to help school children with behavioral problems and learning disabilities.
- *Human resources (HR) psychologists* work in personnel departments to improve hiring practices.
- *Organizational psychologists* help businesses operate more efficiently and effectively.
- *Clinical psychologists* treat people with mental disorders.
- *Counseling psychologists* help people with difficulties in various areas of life, such as marriage, family, and career.

CHAPTER 1

Research Methods

Many disciplines explore human behavior and the human mind: archeology, sociology, anthropology, economics, history, literary criticism, and theatre, to name a few. Even the novelist and the choreographer are students of behavior. The social sciences are distinguished from the other disciplines, such as history, because they employ the scientific method. Psychology is distinguished from the other social sciences because of the specific methods it employs to understand the mind and behavior.

In this chapter, we discuss the scientific method and how psychologists have adapted the scientific method to the study of behavior.

Overview

While psychology uses surveys and field research and other methods, the hallmark of psychological research is the controlled laboratory experiment. Even when psychologists leave the laboratory, for example, to discover what is going on in the clinic or the classroom, they bring their experimental techniques with them.

To whatever degree is possible, psychologists try to bring the virtues (and vices) of laboratory research to bear on all their research questions.

HISTORY

Of course, there is no particular reason for this focus on experimentation. Science employs many methods: observation, experiment, and theory. Astronomy and archeology are sciences based principally on observation. Sociology and anthropology are focused on field research, rather than laboratory research. Psychology's twin focus on experiment and the laboratory derives from the historical trends that led to contemporary psychology.

The work of three different groups of researchers led to the development of modern psychological methods. First the physiologists performed experiments on intact animals, probing them to discover the reactions of the nervous system. From the physiologists, modern psychology derives the notion of an *intervention,* something that the experimenter does to alter the system being studied. An intervention is the way experimenters model *causes* in the laboratory. Then they look for the *effects.* The idea is that, in order to understand a complex system, the causes and effects must be distinguished ahead of time, and cause-effect pairs must be tested separately (or in very small batches). In general, potential causes are manipulated and potential effects are measured.

Second were the introspectionists, the proponents of what was called structuralist psychology. They performed experiments on themselves, training themselves to describe their own internal experience of various events in narrowly specified terms. From the introspectionists, modern psychology derives the notion of *experimental control.* Experimental control means taking various measures to ensure that any events occurring in the laboratory are due to the causes being studied, rather than some random event. What the structuralists discovered was that human behavior was highly variable. Even under circumstances where physical and even biological systems would produce identical results, different people behave differently. Even though the methods of the introspectionists were abandoned, the lesson is clear: Experimental control is much more difficult in psychology than in other sciences.

Finally, there were the statisticians. In the late 1800s, Francis Galton and others began to develop mathematical tools for dealing with large tables of numbers, such as information from a census or a survey. These tables contain different types of information (listed down the columns) about many individuals (listed along the rows). For example, there might be a row for each person, with columns for age, sex, height, weight, and occupation. Psychologists call the types of information corresponding to the column headings, *variables.* The different numbers inside the table cells are called *values.*

Tips on Terms: Correlation Coefficient

The statistic used to measure the correlation is called the *correlation coefficient*. It is a measure that ranges from minus one when two variables are inversely related (one is higher when the other is lower), through zero when there is no relation, up to plus one when the two are perfectly directly related (both are high or low together). In the real world, perfect correlations of one or minus one and the complete absence of correlation (zero) are almost never found. Most correlations are somewhere in between.

Galton's key discovery was the *correlation*. The correlation is a measure of the relation between two sets of values. For example, suppose we had a list of the heights and weights of a large number of people. We would find that, in general, the taller people weigh more. This is known as a positive correlation, when larger values of one variable (height) correspond to larger values of the other variable (weight). Calculating the correlation between height and weight would give a higher number than, for instance, the correlation between weight and IQ. A negative value for correlation indicates that higher numbers for one variable make for lower numbers for the other. For example, the number of school absences might correlate negatively with grade point average.

Now, suppose you believed that childhood nutrition affected IQ. If you had a table listing the number of breakfasts missed and the IQs of a large number of children, you would expect that there would be a negative correlation between those two variables. The more breakfasts a child misses, the lower her or his IQ. In this way, the correlation and other statistics like this can be used to look for possible causal relations. Correlations can identify basic numerical relations between variables that are to be expected if causal relations actually exist. In order to test for these causal relations, more sophisticated statistics are involved.

CURRENT APPROACHES

The birth of the modern era in psychological research methods came in 1920 when William Gossett, writing under the name Student, invented a statistical procedure

Critical Caution

It is very important to understand that measurements of correlation cannot, in and of themselves, tell us whether or not one variable *causes* the other. They only tell us where to look. Psychologists even have a slogan: "Correlation does not imply causation."

called the *t* test. The *t* test takes a measure of an overall difference between the values of two variables and relates that measure to the probability that that difference could be due to chance alone. In other words, the *t* test went beyond a relational measure, such as the correlation, to allow an interpretation of the size of the relationship in terms of probabilities. The stronger the correlation, the bigger the number. The bigger the number, the less likely it happened by accident (assuming that other possibilities have been eliminated by the researcher).

Sir Ronald A. Fisher read Student's paper and realized that two of the methodological problems faced by psychology could be solved by using Student's method. Suppose we think that ginseng tea increases IQ. We take a bunch of people and divide them into two groups. We give ginseng to just one group. Then we measure the change in IQ for every person. Even if ginseng doesn't affect IQ, there will be differences between the two groups. The difference in the changes in IQ for one group over the other might be due to the fact that we are dealing with two groups of different people and all of the variability of IQs. In short, there will always be differences between any two groups due to chance differences in the people and the conditions, and other such factors. The *t* test tells us about the probability that any change in IQ is due to the ginseng or to chance differences in the original IQs of the two groups.

This simultaneously solves the problem of how to evaluate interventions as potential causes and how to control for variability. Possible causes are used for interventions. Different interventions define different groups. The groups not given interventions, called *control groups*, provide a measure of the behavioral variability. The statistical test takes all of this into account and gives back a number that can be used as a criterion for asserting whether or not there is evidence of a cause-effect relationship.

Over the past 80 plus years, the number and variety of research methods and statistical tools in psychology have expanded enormously, but the group experiment, with its interventions and controls and statistical analysis, remains the prototype for the scientific study of behavior and the mind.

What Is Science?

Any science is defined not by the questions it asks, but by how it asks the questions. Thus, science is defined by the methodology it employs. It is a systematic study of natural phenomena in which, through the principles of the scientific method, researchers agree on how to evaluate competing theories. Even though scientists may disagree on which theory is better, they agree on what is considered acceptable evidence by which to evaluate a theory.

THE PURPOSE OF SCIENCE

There are three goals of a scientific psychology:

- **Measure and describe the behavior**   Answers the question: What is the behavior?

- **Understand and predict the behavior**   Answers the question: Why does the behavior occur?

- **Apply the knowledge and control the behavior**   Answers the question: How can the behavior be changed?

The first goal is to measure and describe behavior. The understanding of human behavior must begin with careful descriptions of how people think, feel, and act in specific situations. The second goal is to understand and predict behavior. Psychological explanations are often based on the theoretical perspective of the psychologist, but without this goal, psychology would contain a lot of information with no meaning. Once psychologists understand a behavior, they can make predictions. The third goal is to apply the knowledge gained and control the behavior. In understanding behavior, psychologists are able to exert some control over the behavior, either in the laboratory or in the world.

It is important to know that even though a scientific endeavor may not produce an immediate application, it is still vital to our understanding of how behavior works. This is the distinction between *basic* and *applied* science. Basic science is science for the sake of understanding. The goal of basic science is to understand behavior without necessarily focusing on how that understanding can be used. Applied science, on the other hand, is science for the sake of problem-solving. The goal of applied science is to solve specific problems that exist in our environment.

Key Point: Basic Science

Many people do not understand the importance of basic science. They assume that if the result of the scientific endeavor does not have any immediate application to make the world a better place then it is a waste of energy and resources. However, this neglects the very real fact that without basic science, scientists would not understand a phenomenon enough to be able to apply it to a real-world situation. Hence, the saying, "Without basic science, there would be no science to apply."

Reading Rules

The process described next represents an ideal. In the real world, the steps described here don't always occur in the right order and not all of the steps are performed all of the time or precisely as described.

HOW TO DO SCIENCE

Methods are the how-tos of getting things done. In this section, we will take a look at how science is done.

Theories

Psychologists begin with a *theory,* a comprehensive explanation of a natural phenomenon that predicts a lot of observations based on a few assumptions. Theories reduce the amount of information needed to allow scientists to infer observations logically in terms of just a few basic notions. Thus, a theory is much more general than the observations it attempts to explain. For example, some people accept the notion that the position of celestial bodies has an influence on human behavior. Astrology is a theory in that it predicts a lot of observations about behavior based on an assumption about a relationship between the position of celestial bodies and behavior.

Part of the job of a scientist is to evaluate each theory in terms of the existing evidence. To do this, scientists use three criteria to determine what a good theory is:

- **Fit the known facts** A good theory is compatible with the evidence at hand.

- **Make predictions** A good theory is able to generate predictions of what will happen in the future.

- **Be falsifiable** There must be some way to disprove the theory. A theory that predicts that everything will happen is predicting nothing at all.

Critical Caution: Science Never Proves Anything

A theory can never be proven, and scientists rarely use the word. A theory can be supported or disconfirmed by the evidence, and confidence in the theory can either increase or decrease, but we can never prove that a theory is correct. This is because something outside of the theory, which we have not yet thought of, may also explain our observations. You can support a theory, or disconfirm a theory, but you cannot prove a theory correct.

To continue our astrology example, if the available data suggest that there is no relationship between the phase of the moon and abnormal behavior, then the theory that the configuration of celestial bodies affects human behavior would not fit the known facts. Second, our theory of astrology is not really useful for understanding human behavior if it does not allow us to make predictions about future behavior. Finally, there must be some way to disconfirm the theory, or demonstrate that the theory is incorrect. For example, if our theory of astrology states that the configuration of celestial bodies causes people to behave, it will always be true, because people are always behaving in some way or another. To be a good theory, astrology must predict that, in some situations, a certain behavior will occur and another behavior will not.

What happens when two competing theories fit the known facts, make predictions, and are falsifiable? Scientists use the principle of *parsimony*. All else being equal, the better theory is the theory that explains reality in the simplest terms using the fewest number of assumptions. For example, let's assume, for a moment, that our theory of astrology fits all of the criteria of a good theory. We can find an equally good theory of human behavior that states that abnormalities in brain chemistry affect behavior. The latter theory is more parsimonious because, in the case of brain chemistry, we don't have to assume a connection between the causes and their effects on behavior. We already know that behavior involves muscular action and that nerves connect the brain to the muscles. In the case of astrology, we would need to add an extra assumption as to *how* celestial bodies affect behavior.

Hypotheses

Once we have a good theory, we use that theory to derive *hypotheses*. A hypothesis is a tentative statement about what happens in a given situation. It is the testable prediction, derived from theory. We might derive a prediction from our astrological theory that people will behave abnormally when the moon is full. From this prediction, we can construct a more formal hypothesis that states that people will exhibit more abnormal behavior when the moon is full than when it is not.

Tips on Terms: Constructs

In psychology, theories are often expressed in terms of causal relations between broad conceptual categories. In our example, "positions of celestial bodies" and "abnormal behavior" are *constructs* used in the theory.

Designing Studies

Once we have decided what theory we are working with and have derived a testable hypothesis, the next steps are to design the research study and to gather evidence to support or disconfirm the hypothesis. To do this, we design the study, gather the data, analyze the results, and interpret the meaning of the results in terms of our theory.

The *method* is the procedure scientists use to test hypotheses. It varies considerably based on what hypothesis is being tested. Basically, the researcher is trying to either find or develop the best way of objectively and accurately measuring the behavior of interest. We will discuss specific types of research designs later in this chapter.

Returning to our astrology example, we first need to define what abnormal behavior means and determine the best way to measure it accurately. We may decide that it means a higher crime rate and collect police reports as a measure. We may alternatively decide that abnormal behavior is streaking through the streets of town and measure it by sitting at an intersection and counting the number of naked people running through the streets. However we decide to measure the behavior of interest as part of this study, it is important to measure it the same way each time. In other words, once we have developed a methodology, it must be *specific* and we must follow it precisely each time.

Gathering Evidence

In science, gathering evidence means *measurement*. Whether we are doing observations in the field or in the laboratory, we observe what is going on and then record it. The recording, whether on paper or video or computer, whether in the form of numbers or words or images or sounds, is a measurement of what happened. As we will see later in the chapter, it must be the case that all of the measurements are repeatable. When the same thing happens in the world, the same sort of record must be made. Otherwise, the record won't correspond to what happened, which means

Key Point

The heart of science is *replicability*. Because science is all about what is true or false independent of people's opinions, it must be the case that any evidence one person collects will show the same results if some other person collects it. The specific procedure or method used in collecting the data must be written down so that anyone following the method will do the same things in the same way. Method is what makes science independent of people's personal agenda.

it won't be evidence of what happened. It is the data gathering and measurement procedures that make the data gathering and measurement repeatable and also ensure that the record corresponds to the facts.

Analysis

Once the behavior of interest has been measured, the researcher uses statistics to determine whether or not the effect was meaningful. The *data* are the numbers (or other symbols) produced by following the method, and the *results* are produced by the statistical calculations. Once we have our data, we can use statistics to determine whether there were more streakers on the nights when the moon was full than when the moon was in some other phase. The statistics will also tell us whether the difference, if any, was due to chance or if it was a meaningful difference.

Once the data have been analyzed and the results obtained, the researcher needs to determine what those results mean in terms of the hypothesis and the theory. If, for example, our results indicate that there are more streakers during a full moon than there are when the moon is not full, then the results support our hypothesis.

Before we draw the conclusion that the full moon affected abnormal behavior, we need to look for *alternative explanations*. It might be that the temperature was nice that night, so the streakers decided to come out, regardless of the phase of the moon. It might also be that streakers like to "do their thing" by the light of the moon because it is easier to see and to be seen. In either case, a researcher would have to design a follow-up study that incorporates, and hopefully eliminates, the alternatives into the design.

If the results do not support the hypothesis, all is not lost. We can often learn as much from contradictory results as we can from a study that "works." Typically, instead of throwing out the hypothesis entirely and starting over, the hypothesis can be modified to incorporate the lack of findings. A new study would be designed to test the modified hypothesis. For example, if we found that there was more streaking when the moon was not full, we might modify our hypothesis to state that it may not be the fullness of the moon that affects abnormal behavior, but the amount of light the moon emits. Alternatively, we might decide to examine a different type of abnormal behavior.

How Do Psychologists Do Science?

We have seen how science in general works. Now we will take a look at the ways psychologists have dealt with the specific challenges of studying human behavior in adapting the scientific method to that study.

In broad outline, compared to other sciences, psychology has very informal theories with highly formalized methods. Theories can be expressed in any language and talk about all sorts of processes. In contrast to the harder sciences, new theoretical terms can be added at any time. There are no restrictions on the logic or mathematics used.

On the other hand, methods in psychology, particularly experimental methods, are much stricter than they are in the harder sciences, such as physics, chemistry, and biology. In those sciences, an experiment is just a matter of setting up a situation that challenges the claims of the theory. In psychology, there are highly specific rules for what makes an experiment.

In part, these very formal methods are necessary due to the variability of behavior and the complexity of the human brain. In physics, and even in biology, systems placed under controlled conditions of temperature, pressure, humidity, and so on, behave in very similar fashion. In psychology, simply setting up the initial physical circumstances identically does little to limit the variability seen between one person and another. Complex experimental design and statistics are needed to allow us to see through all this variability in the causes and effects that underlie behavior.

DEFINING VARIABLES

One of the most important aspects of psychological research involves defining the variables that you want to study. *Variables* are measures of the things that may change, either causes or effects, over the course of the research. In our previous example, both the phases of the moon and the number of streakers are variables. The notion of a variable in research goes back to the tables of numbers used by statisticians before the advent of modern psychology. In psychology and other social sciences, hypotheses are very narrowly expressed in terms of changes in the values of variables.

Operational Definitions

In carefully and precisely defining our variables, we choose exactly what we want our terms to mean. This *operational definition* of each variable specifies the procedures used to produce or measure a variable. We operationally defined abnormal behavior in our previous example as streaking. We could also operationally define abnormal behavior as criminal activity, the need for psychiatric patients to increase their medication in order to maintain stable behavior, or the number of times your slovenly roommate cleans the apartment. The only limits of an operational definition are the researcher's imagination and the fact that it needs to make some kind of sense. By operationally defining our variables, we can avoid the problem of another researcher having to guess what we mean by our terminology.

Tips on Terms: The Bell Curve

While we will not be dealing extensively with the mathematical side of statistics here, we will take a moment to discuss the famous *bell curve*. The bell curve is famous because the values of many, many, many variables, when graphed above the number line, form a bell curve or something close to it. There are even mathematical proofs that, given a large enough sample from a population of almost anything, a specific bell curve called the *Normal distribution* will result. So, just what is a Normal curve?

Take any sort of variable—height, weight, intelligence, distances between cities, car prices, shirt sizes, to name a few—and examine all the values for a large number of items. First, there will be a lowest value (called the *minimum*) and a highest value (called the *maximum*). (For instance, no matter how many people we select, there is always a tallest and a shortest person.) If we make a graph showing how many items there are for each possible value, we get what is called a bar chart or histogram. The histogram shows how the values are distributed across the number line, which is why statisticians refer to the contents of a histogram as a *distribution*. The Normal distribution is a particularly useful mathematical function that is shaped like a bell and describes many different kinds of data.

For most kinds of variables, that histogram will be shaped like a bell. (See Figure 1-1). There will be a lot more middling values than there are high values or low values. (There are lots more folks of middle height than folks who are very tall or very short.) The farther you get from the middle values, whatever they are, the fewer items you find.

Statisticians have special names for statistics that measure the bell curve.

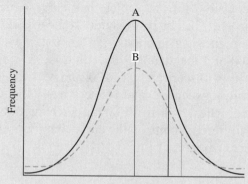

Figure 1-1   Two bell curves with different standard deviations

continued ...

The number on the number line beneath the highest point of the bell represents both the most common value and the value found in the middle of the values. Statisticians call this the *mean,* and it is the same as the average value, which you probably already know how to calculate. The mean locates the center of the bell curve on the number line. (In Figure 1-1, the mean of both curves is marked with the tall vertical line in the center.)

A second important statistic is the *standard deviation*. The standard deviation tells us how fat or skinny the bell of the bell curve is. This gives us a better idea of how spread out the numbers are than if we knew only the maximum and the minimum values.

For example, suppose we are interested in the weights of noncommercial vehicles driving on the road at any given time. The lightest vehicles are the motorcycles. The heaviest are the RVs. Now, let's compare the distribution of vehicle weights for Saturday at 5:30 PM to the distribution for Wednesday at that same time. In both cases, there are some motorcycles and some RVs on the road, so the minimum and maximum values do not change. However, on Wednesday, there are lots more sedans on the road because a lot of folks are coming home from work, and the motorcycles, SUVs, and RVs are parked in the driveway. The bell curve for Wednesday will be taller and thinner than the bell curve for Saturday, which will be more flattened out.

Figure 1-1 shows two bell curves with different standard deviations. The standard deviation is a fixed length along the number line that picks out different proportions of the bell curve as it moves away from the center. We have marked both curves at the points one standard deviation above the mean. For a tall, skinny curve, the standard deviation is smaller, because we don't have to go as far from the center to cover the same number of items; skinny bell curves have more stuff packed into the center.

The trick to the standard deviation is that it is measured in the same units as the original variable. The standard deviation for our vehicle bell curve is measured in pounds. The standard deviation for people's height is measured in inches. The standard deviation for intelligence is measured in intelligence points, and so on. If we are told that the mean for an intelligence test is 100 and the standard deviation is 15, we know that a person of exactly average intelligence will score an IQ of 100, and we know exactly what percentage of the population will score between 100 and 115, between 115 and 130, between 130 and 145, and so on. (Of course, these will be different percentages because the divisions by standard deviation are at different points on the curve, but for each location, we know the percentage for one standard deviation.)

Causal Hypotheses

In the sciences generally, a hypothesis is any fact that can be tested by experiment or observation. In psychology, hypotheses are considered much more narrowly as the specific causal relations predicted by a theory. For example, P causes Q, X

Critical Caution

Operational definitions are extremely useful in a highly diverse science such as psychology that lacks a single vocabulary or basic theory. However, there are dangers to operational definitions as well. The main problem is the risk to generality. Suppose we demonstrate that the phases of the moon affect streaking. In order to demonstrate that the phases of the moon affect abnormal behavior more generally, we would have to perform additional experiments testing for other things such as criminal activity, psychiatric medications, and room cleaning. Only after a sufficiently wide variety of effects have been demonstrated could we argue that there is a general relationship between the phases of the moon and the general construct of abnormal behavior. As we mentioned earlier, it is a critical aspect of theories that they be general. Therefore, the use of operational definitions (called *operationalization*) calls for additional experiments to achieve the required level of generality to support theories.

causes Y, etc. If two competing theories differ in terms of something other than a causal relation, it is very difficult, perhaps impossible, to test the difference using standard psychological research methods.

There are two categories of variables used in specifying research hypotheses. *Independent variables* measure possible causes. *Dependent variables* measure possible effects.

Independent variables are the variables that are manipulated or controlled in a study by the researcher. It is the factor in the environment that the researcher hypothesizes to be the cause affecting the behavior of interest. In the previous example, the phase of the moon is the independent variable. Manipulating the independent variable allows a researcher to define groups of people to compare. We can compare the number of streakers on nights with a full moon to the number of streakers on nights with a half moon. Thus, the phase of the moon, full versus half, allows us to define our groups. This is a common technique for operationally defining the independent variable.

Tips on Terms: Independent and Dependent Variable

People often struggle with the distinction between the independent and dependent variable. An easy way to remember the difference is to remember that the terms *dependent variable* and *dependent measure* are interchangeable. Thus, the dependent variable is the variable that is measured.

Fun Fact: Physics Envy

It is no accident that, in psychology, hypotheses are defined in terms of causal relations. Most of the statistical techniques require it. Sir Ronald A. Fisher, who developed the notion of experimental design in psychology, believed that all scientific advances were due to the resolution of disputes between theories over differences about causal relations using targeted experiments called *crucial experiments*. Philosophers no longer take the notion of crucial experiments seriously. Almost everyone agrees that there are other ways for science to make progress. However, the very strict rule for psychological methodology makes it very hard to do other sorts of research in psychology.

Intriguingly, even back in the 1920s, when Fisher put forward his ideas about experimental hypotheses, there were already famous experiments that did not fit in with the notions of independent and dependent variables. Physicists like Millikan and Rutherford had already received Nobel prizes for experiments that could not be done following the strict methodological guidelines used in psychology today.

Dependent variables are the variables that are observed by the researcher. In psychological research, it is the behavior of interest. Just as we operationally define abnormal behavior as streaking, we operationally define streaking as the number of naked people running down the street on that particular evening. Therefore, a dependent variable is operationally defined by how it is measured.

SAMPLING

When chemists want to find out something about copper, they don't start with just any old piece of copper. Different pieces of copper will have different impurities that may create differences in the laboratory. To ensure that the study results can be generalized to all copper, the chemist starts out with specially prepared copper that is "chemically pure." In psychology, there are no psychologically pure subjects. Instead, psychologists pick their subjects from the general population using something called *sampling*. Sampling is designed to ensure that research results found with the subjects of the study apply to a more general population.

It would be impossible to have everyone in the population of interest participate in a research study, so psychologists use various sampling techniques. *Sampling* refers to the process of narrowing down the field of participants. The *population* refers to the entire population of interest, which may be every adult, every adult in the United States, all females attending college in Oklahoma, or every person between the ages of 12 and 24 who suffers from depression, and other such syndromes. From our population of interest, we take a sample of that population who will participate in our research. Then, we are able to generalize from the sample to the population of interest.

There are several types of samples, each one with its own advantages and disadvantages. A *representative sample* is one in which the important subgroups of the population are represented in the sample according to their percentages in the population. For example, a researcher wants to know how men and women differ on several issues. Here, the population of interest is all adults in the US, and that population consists of 51 percent females and 49 percent males. The researcher wants to sample 100 people, so she recruits 51 females and 49 males.

It is important to note that representative sampling is typically very complex. For example, a researcher wants to conduct research on voting preferences in a national election. What other demographic variables in the general population might be important? The sampling will probably want to reflect political party affiliation, age, area of the country, socioeconomic status, religious preference, and a host of other variables. Given a large number of variables to consider, the researcher may have to search for a single 39-year-old middle-class female in North Carolina who is a member of the Green Party and practices Hinduism to represent that 1 percent of the population. Thus, the more variables you introduce into your sampling criteria, the more difficult it will be to find the individuals who are representative of each variable. However, by not including enough sampling criteria, you run the risk of ignoring important variables.

In most cases, a researcher does not need to have a truly representative sample in the sense of capturing the correct proportion of every value of every variable that might be important. The characteristics of the population may not be relevant to the question at hand. In these cases, a *random sample* will suffice. In a random sample, every person in the population of interest has an equal chance of being included in the sample. This type of sampling takes advantage of the assumption that if everyone has an equal chance of participating in the study, any differences between the population and the sample will be evenly distributed between the two.

In many cases, it is not possible to sample randomly or representatively from the population. There are some types of research that require only warm bodies in the laboratory. This is commonly known as a *convenience sample*. Many of the large research universities use convenience sampling from the university population. Often, introductory psychology students are required to participate in psychological experiments for course credit. In these sorts of experiments, we settle for a convenience sample, knowing that it may not do as good a job as a truly representative sample in terms of telling us about the population we actually care about.

In other cases, we may have very good reasons for sampling from a different population from the one we care about. Experiments that cannot be done safely or practically with humans may be done with other sorts of animals. Researchers using nonhuman animals do not need to rigorously sample from the rat population or the pigeon population. Indeed, it is better that they are assured that their research subjects are genetically homogeneous and thus do not represent the population at large. Here representativeness is sacrificed in order to control for variables that are deemed irrelevant to the study. (Genetic variations in the rat population won't help us understand how rat behavior relates to human behavior.)

Issues in Psychological Research

Each issue discussed here is an important part of developing the research methodology used to test a hypothesis.

RELIABILITY AND VALIDITY

Science attempts to say things about the world based on evidence. The scientific method involves all of the things scientists must do to be as sure as possible that what they say is correct. Psychologists use the terms *reliability* and *validity* to distinguish between two sorts of problems that can arise when we try to say things based on evidence. Something is reliable when random occurrences are unlikely to make what we say incorrect on any given occasion. Something is valid when the evidence tells us what we say it tells us about the world.

Psychologists worry about reliability and validity of measurement and of results.

Measurements

Every research study must use measurements that correctly reflect what actually goes on. This applies to both the independent variables and dependent variables. Psychologists have identified two criteria of good measurement: *reliability* and *validity*.

A measurement is reliable if we consistently get the same results every time the variable is measured. Measurements can differ over time, persons, or situations. If the only difference between two measurements is when they are taken, then we are measuring the same person's behavior in the same situation, and, ideally the result should be exactly the same. If either the person or the situation or both changes, then the results should change only due to those changes. A measure of IQ should change with person, but not with situation. A measure of eyeblink should change with the situation, but not with the person. A measure of blood pressure should change with both the person and the situation.

Validity refers to whether our variable is measuring what it is supposed to measure. Again, this applies to both dependent and independent variables. The measurement of an independent variable is valid if it is creating the groups we want it to create. If we want to study how depression affects opinion, we want a valid method for determining who is depressed and who is not. We would not give subjects an anxiety assessment to divide them into groups. The anxiety assessment may be a valid way of determining who is anxious, but not who is depressed. For a dependent variable, we want to ensure that we are measuring the behavior we care about. If we want to study the effects of drugs on depression, we need to make sure we are measuring changes in depression, not anxiety.

In the case of an independent variable, we often have some practical control of validity because we are measuring our own intervention. If we want to study how loud noises affect opinions, we can be pretty sure that our subjects will be hearing noise (and not something else), because we will be making the noise. In the case of a dependent variable, measurement is what determines our results, what effect we will claim is due to the cause. Here, it is especially critical that our variable measures what we say it measures. Otherwise, we might get a powerful effect in our experiment and be entirely wrong about what it is we have discovered. If we are measuring opinions with a questionnaire, we had better be sure that the answers actually reflect the subjects' opinions.

There are several types of validity. (All apply whether it is an independent or a dependent variable that is being measured.)

- **Face validity** The measure appears to assess the variable of interest from the subjects' perspective. A depression survey has face validity if the questions are related to being sad.

- **Content validity** The measure covers all aspects of the variable of interest from the researcher's perspective. If the depression survey asks questions about all of the major clinical symptoms of depression, then it has content validity.

- **Criterion validity** The measure yields comparable results to a different, already validated measure. If scores on a new depression survey match those on an already established scale, then it has criterion validity.

- **Construct validity** The measure yields results consistent with other things known about the associated construct. If a theory predicts that depression has a 30 percent co-morbidity rate with social anxiety disorder, the new survey should be able to pick this up as well.

Results

Sometimes, psychologists also speak about reliability and validity of results, particularly in the case of experiments. A study is said to be reliable if it is both *repeatable* (that is, it is done in a way that others can do it and documented in a way that others know how to do it) and *replicable* (that is, if it is done again, even by someone else, the results will be the same). A study is said to be valid if the results are about the relationship between variables we say they are about and, also, if that relationship applies to the general population we say it applies to (not just our study sample).

PROBLEMS AND BIASES

Psychologists have special concerns with their research that other sciences do not have to contend with. Because psychologists are dealing with living beings that, in most cases, are aware of what is happening around them, researchers have to worry about how this affects research. In this section, we will discuss the common problems in conducting scientific research as it relates to psychology, as well as the unique problems faced by psychologists.

Replication

One problem common to all sciences is how much confidence should be placed in the results of a single study. Even though we might have a piece of research that supports our hypothesis, the result needs to be repeated before we have confidence in it. This is the process of *replication*. We have more confidence in an effect or result if we, or someone else, can repeat the experiment and obtain the same pattern of results.

The simplest way to replicate a study is to do exactly what the original researcher did. This is an *exact replication*. You would operationally define the variables in the same way, use the same dependent variables, use the same script and subject instructions, and so on. The goal of this type of replication is to determine whether or not the effect was due to chance. In the hard sciences, exact replications are common. Different laboratories double-check the newest results from other laboratories before they move on to the next study. In the case of less important results, replication studies are given to students and new researchers as exercises.

Exact replications are all well and good, but science would be rather boring if all we did were exact replications. In psychology, most replication is done through *conceptual replication*. This method of replication attempts to replicate the original study using conceptually similar, but methodologically different, procedures. To conduct a conceptual replication, a researcher will use different manipulations of

> ## Big Background: Replications in Psychology
>
> There are two reasons why conceptual replications are more common in psychology than exact replications. The first is that group experiments with group statistics actually incorporate some of the effects of replication of single experiments in the hard sciences. The second is that, as we mentioned earlier, in order to evaluate a theory, the constructs must be tested using various different operational definitions in order to ensure generality. The use of conceptual replications in psychology is a good thing. The shortage of exact replications is not.

the independent variable or different measures of the behavior of interest. The goal is to determine whether the result is due to the methodology and materials or the concepts they represent.

Confounding Variables

Confounding variables, or confounds, are a problem with all scientific research. Confounds are variables that the experimenter has not controlled for, that might still affect the results. Confounds are a problem in that we can't be certain whether the result is due to the manipulation of the independent variable or to the confounding variable. Researchers often deal with confounds by randomly assigning subjects to groups, but they must also be clear about the effects (and side-effects) of the manipulation of the independent variables. Using the astrology research example from earlier in the chapter, the amount of light emitted from the full moon is confounded with the phase of the moon. There is no way to tease out the effect of the position of the moon with the effect of the amount of light the moon emits. The confounding variable could account for any differences between the groups. Sometimes getting rid of confounding variables is a matter of clever experimental design. (We could measure abnormal behavior that occurred indoors with artificial light.) Sometimes, however, there is no way to unconfound the variables.

Demand Characteristics

Demand characteristics are the cues inherent in the experimental situation that can give a subject a clue as to what is expected of them. They are characteristics of the experimental setting that "demand" a certain behavior. Something about what the subjects are exposed to leads them to believe that a certain behavior or response is expected. For example, suppose we spent the night at a police station and asked each person under arrest, "Did you feel any strange influence from tonight's full moon?"

Obviously, if a subject feels that they are to behave in a certain way, the results will be biased. You will not be able to determine whether your results are due to your independent variable or the demand characteristics of the situation. This is particularly problematic for research using human subjects, but animal subjects can fall prey to this as well. A well-designed study and a quick check around the laboratory setting can easily take care of this problem.

Experimenter Bias

The experimenter can also introduce bias into a research study. An experimenter typically knows what the hypothesis is, which experimental group the subject has been assigned to, and what behaviors are expected. Armed with this knowledge, the experimenter can unintentionally and subtly communicate these expectations to the subjects, which can bias the results. Even if they don't communicate anything to the subjects, the experimenters' expectations can affect their own behavior and bias their observations.

There are two methods that researchers use to avoid this bias. First, they standardize the methodology. By making the instructions and methodology as consistent as possible across each of the experimental sessions, the bias can be reduced. The best way to do this is to deliver instructions and materials on video or audio tape, so they are the same each time. A second way to reduce bias is to use *blinded* studies. A blind study is one in which even the researcher does not know what group the subject has been assigned to. The logic is that if the experimenter does not know which group the subject is in, she cannot know what to expect from the subject.

Subject Bias

The subjects themselves can also introduce bias into scientific research. This typically comes about because the experimental situation is new to most people. Because they are in an unfamiliar environment and do not know what to expect, they often feel uneasy. This can manifest itself in four different ways. "Good subjects" are the subjects who alter their behavior in order to try to confirm what they believe the hypothesis to be. They are trying to help the experimenter. "Faithful subjects" are the subjects who alter their behavior in order to explicitly follow the experimenter's instructions. Generally, following instructions is desired, but there are cases in which it can introduce bias because it is not how the person would normally respond. "Negative subjects" are the subjects who alter their behavior in order to mess up the experimenter's research. Finally, "apprehensive subjects" are the subjects who alter their behavior to look good or be correct because they are worried that the experimenter is judging them personally. The key to each of these subject biases is that the person is altering his or her behavior.

ETHICS

Because psychologists so often deal with living creatures, we have special responsibilities in doing research. When a chemist drops copper into acid, no one worries about the pieces of copper. Even with the changing ethical standards in our society, thus far, when a botanist slices a piece of mold to put it under the microscope, no one worries about the mold.

The American Psychological Association and other professional organizations have put together strict codes of conduct to cover many areas of psychological endeavor, including research. Psychologists are not only responsible for not mistreating their subjects, they must also respect the privacy of those subjects. The most important cases are where the psychologist's responsibility to act ethically conflicts with the need to do good science. In order to design the most effective motorcycle helmets to make people safe, experiments are done where chimpanzees are given head injuries. Do we allow chimpanzees to be injured and killed to protect humans?

There are two important current issues in the ethics of psychological research. First, ethical standards are changing. What was deemed ethical decades ago is now unacceptable. Second, in the United States, the federal government has recently set up an extensive system of review panels (called *Institutional Review Boards,* or *IRBs*) to check the ethicality of all government-sponsored research. Almost all research with human or nonhuman animal subjects must pass review or it will not happen.

Changing Ethical Standards

Many of the most important pieces of psychological research from the 20th century could not be done today, because, by contemporary standards, they are unethical. In some classical cases of unethical research, particularly medical research, people were physically harmed. For example, in the Tuskegee syphilis study, syphilis patients were left untreated even when treatments became available. The unethicality of important psychological studies is mostly of a less egregious form. Our understanding of the role of touch in child-rearing comes from the Harlow monkeys, half of whom were separated from their mothers soon after birth and who suffered severe developmental problems as a result. Our understanding of conformity comes from the Milgram study, where human subjects were fooled into thinking they were delivering noninjurious, but very painful electrical shocks to fake subjects. The harm to Milgram's subjects is hard to quantify, but some of these subjects reported feelings of guilt after the experiment.

That some medical researchers felt it appropriate to deny medical treatment to syphilis patients as recently as 50 years ago probably shocks most of us. That current standards prevent psychologists from doing studies where some proportion of the subjects might feel guilty afterward is a different matter.

Institutional Review Boards (IRBs)

Institutional Review Boards or IRBs were set up in the wake of some of the ethical scandals in medical research. The idea was to ensure that any researcher who accepted government funds, or who worked at or was affiliated with any institution that used any government funds, would have to abide by a set of ethical standards. There is a large network of IRBs throughout the research community. (Most large universities have at least three or four.) Whenever someone wants to do research with human or nonhuman animals or with the records of individual people, he or she must get IRB approval in advance.

The IRB system is very new and is still evolving. Problems have arisen. One of the biggest difficulties is that the system was designed with medical, rather than psychological, research in mind. Elaborate safeguards and lots of paperwork and series of reviews are involved in authorizing studies where blood is drawn, medication is given, surgery is performed, and so on. But the same IRB system set up to address the obvious risks to subjects in these studies must also authorize studies in psychology where the most intrusive thing planned is to give the subjects a questionnaire that asks about how uncomfortable they feel when exam time rolls around. While there are more extensive and less extensive types of reviews for more and less intrusive research, the boundaries are not always clear.

Statistics

Statistics is the study of numbers that are measurements of reality. In psychology, statistics provide the link between the record of the data collected and our understanding of the phenomena based on that data.

GOALS OF RESEARCH STATISTICS

In psychological research, statistics are used to convert the data recorded from a study into mathematical statements that can be used to say whether a hypothesis must be rejected or not. Statistics are also used to get a picture of the data in ways that can lead to new theories and hypotheses.

Exploration

In *exploratory data analysis (EDA)*, statistics are used to get the big picture from the data. In their original form, the data are usually a big table with lots of numbers in it. It is hard to see from just looking at the table what is really going on. What are the patterns to be found in the data? Are values increasing or decreasing? Is this

Tips on Terms: EDA and CDA

In *exploratory data analysis (EDA)*, researchers use statistics to find patterns in the data. In *confirmatory data analysis (CDA)*, researchers use statistics to relate patterns in the data to hypotheses.

variable correlated with that variable? And so on and so forth. Statistics can help by assigning values to specific summary measures such as averages, variances, and correlations among others.

Every statistical measure summarizes multiple values, so every statistical calculation can be used to reduce the number of numbers and help to find the big picture. The trick to EDA is to find the right statistical calculations that will help show the important features of the data. This picture may be shown with summary numbers or with actual pictures, like charts and graphs.

Confirmation

Fisher's strategy for using the *t* test as part of a process for evaluating an experimental hypothesis has become a huge field for using many sorts of statistical procedures to answer research questions from experiments and other types of studies. This process is called *confirmatory data analysis (CDA)*, because we want to confirm the hypothesis based on the patterns found in the data.

Confirmatory data analysis uses a particular type of statistics called *inferential statistics*. The *t* test was the first inferential test. Now there are many others.

Here are some common inferential techniques:

- **Student's t test**   Used to determine if there is an effect on some dependent measure (increase or decrease) due to the experimental group receiving an intervention while the control group does not.

- **Analysis of variance (ANOVA)**   Used with multiple groups for multiple interventions, for example, placebo, low, middle, and high doses of a medication. Also used when there is more than one independent variable. For example, placebo, low, middle, and high doses of a medication (dosage dimension), given to men and women (gender dimension), with and without previous symptoms (health dimension).

- **Multivariate analysis of variance (MANOVA)**   Used to determine the effect of multiple groups on more than one dependent measure. For example, the effect of placebo, low, middle, and high doses of a medication (dosage dimension), given to men and women (gender dimension), with

Tips on Terms: Descriptive and Inferencial Statistics

The various statistical measures and the types of analyses and calculations that lead to them are classified as being either *descriptive* or *inferential*. Descriptive statistics summarize information about the data. Calculations such as sums and averages that provide one or two numbers to tell us about a larger bunch of numbers are descriptive statistics. Inferential statistics are created using much more specialized calculations designed to help make decisions as to whether research data is sufficient to disconfirm a particular hypothesis.

The word *inferential* is used because the calculations help us confirm hypotheses by relating the patterns found in the sample data to the possibility that those same patterns apply to the population from which we sampled. Experimental hypotheses are stated in terms of differences found with the experimental subjects. But what we really care about is the people outside of our laboratory. Inferential statistics allow researchers to say how likely it is that the pattern found in the data will be found in the population. This is called making an *inference* from the sample to the population. (Technically, we could make an inference about a population not related to a hypothesis. This would be a case of using inferential statistics in exploratory data analysis.)

While the purpose of description is often exploratory and the purpose of inference is most often confirmatory, both types of statistics are used in both types of analysis. In CDA, descriptive statistics are a vital preliminary step in establishing the pattern of the data before hypotheses can be evaluated. In EDA, inferential statistics can be used (with care) to examine complex relations between variables.

and without previous symptoms (health dimension) on both cure and number of side-effects.

- **Regression** Used when the degree of the impact of the intervention can be measured as a unique number. Instead of using the independent variable to define groups, unique values are assigned to each subject. For example, predicting improvements in health based on the precise blood levels of medication for each patient.

- **Multiple regression** Used with multiple types of intervention. For example, predicting cholesterol based on amounts of medication, exercise, and fat content in diet.

- **Multiple multivariate regression** Used to predict multiple effects of multiple types of intervention. For example, predicting cholesterol, blood pressure, and angina pain based on amounts of medication, exercise, and fat content in diet.

Big Background: Why Are Statistics Important?

Psychology, like many of the social sciences, makes extensive use of statistics. For those of us who continue on with psychology, there are a lot more statistics ahead. The question is: Why?

Back around 1915, a prominent psychiatrist, Henry A. Cotton, began working on a novel theory of mental disease. Believing that mental disease was caused by toxins given off by bacteria hidden in various places in the body, he began removing impacted wisdom teeth, appendices, tonsils, etc., and even colons and cervixes from mentally disturbed people. In an age before antibiotics, the goal was to eliminate these harmful bacteria.

In 1921, Dr. Cotton gave a series of talks announcing his success. Thousands of people, some seeking relief from mental and emotional problems, some healthy folks seeking preventative treatment, came to his clinic or brought their children. Teeth and organs were removed by the tens of thousands, all to no purpose. The death rate for the abdominal surgeries may have reached as high as 45 percent.

Cotton based his belief that his treatments were effective by using the *case study* method, a vital methodological tool in psychology, medicine, and many other fields. But the case study, by itself, cannot demonstrate conclusively that an intervention is safe and effective. Determining whether a new theory is good or bad requires *inferential statistics*. The inferential statistics developed since 1920 assure us that the changes proposed in our society will help rather than harm us.

These sorts of issues are with us today. Recently, Larry Summers, the president of Harvard University, defended his hiring record by claiming that women were genetically less suited to careers in science and mathematics. While President Summers' field is not in the social sciences, unlike Dr. Cotton, he could point to a good deal of evidence that his claim was true. Unfortunately for him, the notion that women are not as well suited to careers in science and mathematics is still controversial. And what do we mean by *controversial*? We mean the claim has not yet passed statistical muster.

There are many proposals whose safety and efficacy have not yet passed statistical muster: The influence of genetics on intelligence. Violence in the media as a cause of violence in society. The medical risks of silicone breast implants. The negative health effects of dietary cholesterol. The superiority of intrinsic rewards over external rewards in teaching. All these and more provide the basis for proposals from both liberals and conservatives who wish to change how you live your life. All have been validated using case studies and observational studies using *correlational statistics*. None has been sufficiently validated using experiments with inferential statistics. In a world filled with crusaders who want to intervene to improve the world according to their theories, statistics is the best protection we have against the ineffective and unsafe practices of folks like Dr. Cotton.

MODES OF PRESENTATION

Snazzy statistical calculations are of no value unless the results can be presented effectively to an audience. Science is a communal enterprise. Research discoveries are only as valuable as the new directions they create in further research. Statistical results can be presented *numerically, graphically,* or in combination. Numerical and graphical presentations both have different advantages and disadvantages. The choice of which to use should be made carefully.

Types of Research

In this section, we will examine the details of different types of research design.

DESIGNS

There is no such thing as the perfect research design. Psychologists try to use the best design possible for the hypothesis being tested. Each design has its own advantages and disadvantages The various designs allow researchers to draw different kinds of conclusions and interpretations. In the following sections, we describe the five categories of research design. It is important to note that within each category, there is a great deal of variation.

Case Histories

Case histories are a detailed account of the phenomenon being studied. There are some interesting phenomena that occur too rarely to be sampled properly, such as photographic memory or multiple personality disorder. Common or not, when a highly complex phenomenon is first discovered, a detailed account of an individual case can guide researchers to important aspects of what is going on.

In either case, a researcher will take a detailed account of the individual's history in order to discover any unusual factors that might have contributed to the condition. A researcher may also discover what effect the rare phenomenon may have on other behaviors. Case histories not only provide us with a wealth of descriptive information, but they may also suggest important principles underlying the phenomenon.

The disadvantage of this type of research is that it cannot give us any information about the causes of the rare behavior or condition, or how the behavior affects other behaviors. We can uncover ideas about potential causes, but we cannot determine cause with this type of research design. The advantage is that it allows us to begin to study rare conditions or behaviors that are too difficult to study in other ways.

Surveys

Survey research designs are an attempt to estimate the opinions, behaviors, or characteristics of a particular population. This is done by posing questions to a sample of the population. This is usually done through interviews or questionnaires, although other methods, such as examining records, may be used.

   The advantage of this type of research is that it is relatively easy to distribute surveys to a large group of people. Two disadvantages are that the answers to survey questions are very easily biased by the wording of the question and there is no way to perform controlled interventions to find out about causal relations.

Correlational Studies

Correlational studies investigate the relationship between two variables that are not controlled by the researcher. These studies investigate how the occurrence of one variable coincides with the presence of the other variable. What researchers are trying to do with this type of design is to predict the occurrence of one variable with

Critical Caution

It is important to realize that just because there is a relationship between two variables, it does not mean that one variable *causes* the other. *Correlation does not imply causation.* This is due to the fact that the statistic used to analyze correlational data can be calculated for any two variables that can be measured in any order.

   For example, we can measure global temperature over a period of 100 years and also measure instances of piracy over the same time period. When we calculate the statistic, we find a strong negative correlation. As global temperature increases, piracy decreases. Can we say that increasing global temperature causes piracy to decrease? It might be that pirates like to do their thing when the temperature is cooler.

   However, let's restate the correlational relationship between global temperature and piracy. As piracy increases, global temperature decreases. Can we say that piracy causes global temperature to decrease? Likely not. However, logically, the correlation between the two variables is just as good evidence for piracy changing the climate as for the other way around. Therefore, more than correlation is always needed.

the occurrence of the other by measuring the relationship. Surveys are a type of correlational study, but correlational studies may be performed in the clinic or the laboratory when no intervention is used.

For example, clinical psychologists discovered that depression and anxiety disorder tend to coexist. They measured the occurrence of depression in a sample and the occurrence of anxiety disorder in the same sample and found that when depression occurs, so does anxiety disorder. No intervention was used, only controlled observations using specialized tests.

True Experiments

True experiments are used to answer questions about cause and effect. The main advantage of this type of research design is that it allows the researcher to control the experimental variables. Thus, they are able to rule out other influences on the behavior of interest. The logic is that by ruling out all other factors except those being studied, researchers can determine a cause and effect relationship between the factor and the behavior. The disadvantage is that not all factors can be controlled by an experimenter.

There are two criteria that must be met in order for an experiment to be a true experiment. First, the independent variable must be manipulated by the experimenter. This means that the factor that the researcher believes to affect the behavior of interest can be completely controlled by the researcher. In the simplest case, the factor is present for one group of subjects and absent for the other group of subjects.

This simple two-group experiment creates an *experimental group* and a *control/comparison group*. The experimental group is the group that is exposed to the presence, or manipulation, of the independent variable. They are the test group that is hypothesized to show the effect. The control group is the comparison group that does not receive the manipulation of the independent variable. They are exposed to everything that the experimental group is exposed to *except* the manipulation of the independent variable. In making the comparison, we can conclude that because the only difference between the two groups is the presence of the independent variable, any differences on the dependent variable is due to the independent variable.

Second, a true experiment must have random assignment to groups. This means that each person participating in the experiment has an equal chance to be in any group. There will always be factors in an experiment that cannot be controlled. Random assignment allows those factors to be equalized among groups. This prevents any random differences in the behavior of interest from correlating with any other possible causes not controlled by the experimenter.

Quasi-experiments

Sometimes it is not possible to meet the criteria of a true experiment. In psychology, there are a number of interesting variables that we can neither manipulate nor assign randomly. We still, however, want to be able to explore cause-effect relationships. A *quasi-experimental* design is a catchall category for studies that fall somewhere between true experiments and correlational studies.

For example, we may want to look at how harsh parental discipline affects juvenile delinquency. Because of ethical considerations, we cannot manipulate how parents discipline their child. In order to manipulate the independent variable of lenient versus harsh discipline, we would need to find parents who already fall into one of the two categories. Therefore, subjects cannot be randomly assigned to groups.

Or we might want to look at how men and women differ on some behavior. However, we will have a difficult time randomly assigning humans to be either male or female (would *you* volunteer for such a study?). This is true for all demographic variables. It is important to note that even though we are still exploring cause and effect relationships, the conclusions drawn from quasi-experimental designs are not as strong as those that we could draw from a true experiment.

LOCATIONS

Where research is conducted can have an enormous impact on the results. Three common locations for psychological research are the *laboratory,* the *field,* and the *clinic.*

- When a study is conducted in the laboratory, the researcher can control many, many more of the variables and be much more confident as to what causes what. The downside is that the laboratory itself is an unnatural environment. There may be differences in the behavior in the real world.

- *Field* studies are conducted in the real world. Observations may also affect behavior, but this is less likely to occur than when re-creating the behavior in the laboratory. The downside, of course, is the loss of control of so many variables.

- For psychologists, the *clinic* can be used as a compromise between the field and the laboratory. It is a part of the real world where psychologists can exercise a certain amount of control. There are additional ethical concerns, however. If a clinic study performs different treatments on different patients, what are the researchers obligations to the patients receiving the less effective treatment?

APPROACHES

Over the last hundred years or so, psychological research has expanded enormously. There are too many variations in research methodology to count. Here are a few of the more common ones:

- **Cross-cultural**   Psychological phenomena that are part of human nature should be the same in all parts of the world. An effective way to see if a psychological phenomenon has cultural factors is to perform the same study in several different parts of the world to see if the behavior is different in different cultures.

- **Multi-disciplinary**   Many questions of interest to psychologists are also of interest to other researchers. Researchers from other disciplines also have skills useful to psychological research. Psychologists collaborate with researchers from many other disciplines, including medicine, education, computer science, political science, anthropology, sociology, economics, and others.

- **Computer simulation**   When a theory is highly complex, it may not be obvious exactly how the behavior would look if the theory is true. In other words, we don't know what hypotheses can be derived from the theory. Computer simulations of theories can calculate the effects of a theory and help us create hypotheses.

- **Meta-analysis**   Meta-analysis is a set of statistical techniques designed to compare and combine the results of multiple studies. When there are multiple conceptual replications with differing results, meta-analysis can give us a bottom line for the studies combined.

Summary

Research methods in psychology, particularly those governing experiments, tend to be stricter than in other sciences. As a result, in order to enable a variety of methods, research methods in psychology have been developed to a greater degree than other sciences. Due to the variability and difficulty in controlling human behavior, psychological methods rely heavily on statistics. The focus of research methods in psychology is to establish causal relations governing behavioral phenomena in order to advance the study of psychology.

Quiz

1. Which of the following is a goal of a scientific psychology?

    (a) Measure and describe behavior

    (b) Understand and predict behavior

    (c) Apply the knowledge and control behavior

    (d) All of the above

2. A(n) _____ is a comprehensive explanation of a natural phenomenon that predicts a lot of observations based on a few assumptions.

    (a) Hypothesis

    (b) Theory

    (c) Control group

    (d) Guess

3. The _____ is the procedure scientists use to test hypotheses.

    (a) Scientific method

    (b) Theory

    (c) Experimental group

    (d) Random sampling

4. The _____ variable is the variable that is manipulated, and the _____ variable is the variable that is measured.

    (a) Dependent; confounding

    (b) Independent; confounding

    (c) Independent; dependent

    (d) Dependent; independent

5. A _____ is one in which every person in the population has an equal chance of participating in the study.

    (a) Representative sample

    (b) Random assignment

    (c) Convenience sample

    (d) Random sample

6. A measurement is _____ if it consistently yields the same results every time and is _____ if it does what it is supposed to do.

    (a) Valid; valid

    (b) Valid; reliable

    (c) Reliable; valid

    (d) Reliable; reliable

7. _____ are the cues inherent in the experimental situation that can give the subject a clue as to what is expected of them.

    (a) Subject biases

    (b) Confounds

    (c) Experimenter biases

    (d) Demand characteristics

8. Which type of research design allows scientists to draw cause and effect conclusions?

    (a) Correlational studies

    (b) Case histories

    (c) True experiments

    (d) Surveys

9. When either manipulation of an independent variable or random assignment to groups is not possible, what research design are you conducting?

    (a) Quasi-experimental

    (b) Case history

    (c) Survey

    (d) True experiment

10. Which of the following is not an inferential statistic?

    (a) Student's t-test

    (b) Averages

    (c) ANOVA

    (d) Regression

PART ONE

Actions in the World

We begin our survey of topics in psychology with an examination of those areas that address the person-as-a-whole, rather than any specific types of things that people do. Psychology views people from a number of different perspectives. Part One examines the most important of the subfields of psychology that examine the whole person.

Chapter 2, "Social Psychology," covers the study of the person as a social being. Chapter 3, "Developmental Psychology," looks at the study of the person growing and changing throughout life. Chapter 4, "Intelligence," looks into the study of how we differ in terms of our intellectual activities. Chapter 5, "Personality," examines how we differ in other ways. Chapter 6, "Health Psychology," provides an example of how psychology is applied to improve our lives. Chapter 7, "Psychological Disorders," talks about living problems that are psychological in nature. Chapter 8 "Psychotherapy," discusses what psychologists do to help with those problems.

CHAPTER 2

Social Psychology

Human beings are highly social animals. Our behavior is greatly influenced by our social environment. Our behavior is changed by the behavior of others, and our behavior in groups differs from our behavior on our own. The study of the various influences of other people is the realm of social psychology.

Overview

Social psychology has more in common with the other social sciences, particularly anthropology and sociology, than any other branch of psychology, at least in terms of what social psychologists study. There is even a branch of sociology that examines psychological influences on sociological phenomena. (Oddly enough, despite the fact that it is a branch of sociology, it is also called "social psychology." We won't be covering any sociology in this book.) Social psychology can be distinguished from these other fields because of its strong focus on the experimental method. However, both the topics and methods of social psychology emerged gradually as it distinguished itself from these other fields.

In the end, social psychology has come to include a wide variety of research areas. Social psychologists study couples, changes in opinions in members of groups, group decision-making, how one person can persuade another, and many other topics. Because we are such social animals, the processes they have discovered have implications across psychology and throughout life.

HISTORY

Concern with the scientific study of social phenomena began with Auguste Comte (1798–1857). Comte was a true believer in the applicability of the scientific method even to the highly variable and unpredictable phenomena of behavior. He was very much ahead of his time as a skeptic about the value of speculation over observation. The only thing he was not a skeptic about was introspection. He was a proponent.

Comte's grand vision of the social sciences came about in dribs and drabs. However, even early on, important distinctions in how to study social psychology became evident. First, there was the issue of whether to study individuals or groups. The tradition of studying groups, which became modern sociology, began with Durkheim (1858–1917). Tarde (1843–1904) took a contrary view, hoping to understand groups by understanding individuals, which led to the approach used in psychology.

Another issue was whether to base social psychology around experiments. Outside of psychology, the study of social phenomena is often deeply involved with ways of describing behavior verbally. An author examines the behavior in the field and then writes a book about it, developing a vocabulary for organizing the understanding of the social system. This tradition began early. Durkheim, for example, invented the term, *anomie*, to describe the disconnect between personal values and societal values that presaged suicide. The roles of imitation and suggestion in group behavior were also discussed. Early in the twentieth century, Max Weber introduced the notion of *charisma,* which tied group behavior to particular individual characteristics of one person.

CURRENT APPROACHES

It was not until 1924 when Floyd Allport wrote a book advocating experimentation in social psychology that a recognizable social psychology began to emerge. Experiments in social psychology were about how measurable behaviors changed due to the social context. As we will see, social psychology has tended to keep to a tradition of small theories, closely tied to the hypotheses that can be tested experimentally.

Key to Allport's model was the central dependent variable of social psychology, the *attitude* measure. Typically measured on a ratings scale, attitudes come in degrees from negative through neutral to positive. The two advantages to attitude measures are:

- They can be used to evaluate someone's attitudes toward (or against) anything that can be described.
- They are very easy to measure.

Many of the topics we deal with in this book look at the internal psychology of the person. For example, cognitive psychology examines how people process information internally. On the other hand, social psychology is based on a different theoretical assumption. This is the assumption that an individual's psychology does not exist in a vacuum. It is affected by the social context in which the person exists. Social psychology focuses on how the real or perceived presence of others affects a person's thoughts, feelings, and behavior.

Social Cognition: Thinking in the Social Setting

In the wake of the cognitive revolution, social psychologists have found ways of studying the relationship between thinking and social relations. Two of the most influential areas of study are *attribution*, which examines social influences on where we assume the causes of other people's behavior lie, and *persuasion*, which examines how we change each other's minds.

ATTRIBUTION: THINKING ABOUT OTHER INDIVIDUALS

Attribution is the process whereby we assign causes to both our own and others' behavior. We do this in order to add predictability to the situation or interaction. If we have an understanding of the cause of someone's behavior, we may be better able to control and direct the interaction.

Types of Causal Attributions

There are two factors that we examine when attributing causes of behavior. First, we look toward the source of the cause. A cause can be either internal or external to the person behaving. Internal causes are those that stem from the internal characteristics of the person. These would include their attitudes, abilities, or personality traits. External causes are those that stem from the situation or

Table 2-1   Categories of Causal Attributions

		Stability of Cause	
		Unstable	Stable
Source of Cause	**Internal**	*Effort, mood, or fatigue* "I studied hard for this exam."	*Personality trait, ability, intelligence* "I'm smart."
	External	*Luck, chance, opportunity* "I was lucky this time."	*Task difficulty* "The professor is easy."

the environment. Second, we also try to determine the stability of the cause. A cause can be either stable or unstable over time. Stable causes are those that are enduring over time. Unstable causes are those that are either temporary or vary over time.

With these two factors, we can categorize all potential causes of behavior as falling into one of four discrete categories. Table 2-1 illustrates these four categories using the example of possible causal attributions for your passing an exam. Within each of the four cells in italics is a general category of potential causes. Underneath each category is a quoted statement detailing a potential specific cause for passing the exam.

Errors and Biases

While the process of making causal attributions is interesting in and of itself, psychologists are also interested in how accurate those attributions are. Research has long shown that we are not always accurate when attributing the causes of behavior. Moreover, people tend to make the same kinds of errors or are biased to make a certain type of attribution. We will discuss three types of errors and biases.

The *fundamental attribution error* is the tendency to prefer internal attributions over external attributions. For example, someone you just met is rude to you at a party. According to the fundamental attribution error, you are more likely to attribute their rudeness to an internal characteristic than to an external one. Thus, you are more likely to attribute their behavior to their being an unpleasant person than you are to attribute it to their just having a bad day.

The *actor-observer bias* is the tendency to make the fundamental attribution error more often when trying to explain others' behavior than when explaining our own. When we're the actor, we have a lesser tendency to look to internal causes for the behavior than when we observe the same behavior in another person. For example, you and another person are walking along a stretch of sidewalk on your way to class. You both trip and drop your books. You will attribute the cause of the other person's tripping to their clumsiness (internal), while attributing your tripping to a tiny crack in the sidewalk (external).

Why do we tend to look away from internal causes for our own behavior? First, we have more information about ourselves than we have of others. We know how our behavior varies over situations, so we look more toward external causes. Second, we have a tendency to attribute odd or unexpected behavior to internal causes. Because our own behavior is rarely surprising to us, we look more toward external causes.

The *motivational biases* occur when something motivates us internally to make a particular attribution. We can distort the attributions we make about our behavior in order to protect our ego, enhance our self-esteem, or look good to others. The *self-serving bias* is the tendency to attribute our successes to internal causes and our failures to external causes. For example, if you pass an exam, you will probably attribute your success to your intelligence rather than to the exam being easy. However, if you fail the exam, you will probably look for an external reason for your performance. Thus, your roommate kept you up too late, or the exam was unfair. We will do almost anything to protect ourselves from attributing the cause of our failure to our potential lack of intelligence, which is damaging to our self-esteem.

In some cases, we can take this a step further by creating *self-handicapping strategies*. Here, we create new potential external causes for our behavior in case we fail. We try to protect ourselves from unflattering internal attributions by creating external causes for our failure. For example, you do not think you will do well on an exam. Instead of having to take responsibility for the cause that may lead to a "less intelligent" or "lazy" causal attribution, you stay out all night before the exam. Thus, when you fail, you can attribute its cause to having been out all night. What happens when we self-handicap and we do not fail? We tend to make a stronger internal attribution than we would have if we had not self-handicapped at all. Instead of attributing the cause of our success to "I'm intelligent," we tend to strengthen that attribution to "I'm the smartest person in the history of the class."

ATTITUDES AND PERSUASION: THINKING ABOUT THE SOCIAL SPHERE

A great deal of our time and effort is devoted to determining what opinions or attitudes people hold. You cannot turn on the news, open a newspaper, see a television advertisement, or surf the web without running across either an opinion survey or the results of an opinion survey. Social psychologists do spend some time evaluating what opinions or attitudes people have, but they are more interested in how those opinions came about and how to change them. This is the distinction between attitudes and persuasion. Social psychologists who study attitude change and persuasion focus on the processes behind how we change our minds. When examining what changes a person's mind, persuasion researchers study factors

Tips on Terms: Attitudes and Persuasion

Attitudes are positive or negative tendencies toward people, events, actions, practices, and so on, called attitude objects. Attitudes can usually be expressed by a degree of agreement with a sentence about the attitude object. Attitudes can and do change over time due to many things, including individual experience. When someone else in our social environment changes our attitude, that is called *persuasion*.

related to the persuasive message, the source of the message, and the characteristics of the message's audience.

Message Factors

Regardless of the source or the audience, some messages are more persuasive than others:

- **Position advocated**   The amount of discrepancy between the message and the audience's opinion. A proattitudinal message advocates a position that the audience is already favorable toward. A counterattitudinal message advocates a position that the audience is negative toward.

- **Message sidedness**   Should only one side of the argument be presented or should the message contain both sides? A one-sided message is persuasive if the audience already agrees and the source's job is to make them agree more. A two-sided message is used when the audience is knowledgeable about the topic or disagrees with the position.

- **Repetition**   Should the message be repeated? The mere-exposure effect shows us that repeated exposure to a persuasive message may increase acceptance, but only up to a point. Repetition is particularly effective if the audience likes the item, person, product, or argument.

Source Factors

In some cases, the source of a persuasive message is more influential than the message itself:

- **Source credibility**   This pertains to the believability of the source. People have a tendency to be persuaded by people who are believed to be highly credible with respect to what they are talking about. This credibility can also be thought of in terms of expertise or trustworthiness.

- **Source attractiveness**   More attractive sources tend to be more persuasive than unattractive sources. This can be thought of in terms of physical attractiveness, pleasantness, similarity to the audience, or likability.

- **Power**   The source may have the power to punish the audience if they disagree. This results in increased reported persuasion, but not long-lasting persuasion.

Audience Factors

The characteristics of the audience of a persuasive message also influence the degree of persuasion:

- **Trusting people**   Since they have learned that they can rely on the word of others, people who trust others easily may be more likely to be affected by a persuasive message. This is particularly true if the source is highly credible.

- **Self-esteem**   People with high self-esteem may be less likely to be persuaded because they have more confidence in their own opinions.

- **Forewarning**   Knowing in advance that someone is going to try to persuade you can lead to decreased persuasion. You have the chance to develop counterarguments in advance and steel yourself against the persuasion attempt.

- **Need for cognition**   People differ in their need to think about the arguments in a persuasive message. Those who tend to think more about the message tend to be harder to persuade.

- **Intelligence**   Higher intelligence may increase the ability to understand a persuasive message, but it may decrease the likelihood of accepting the message.

Tips on Terms: Cognitive Dissonance

Thus far, we have discussed persuasion in terms of a deliberate attempt to change attitudes. There are, however, types of attitude change that occur without this deliberate influence. *Cognitive dissonance* describes one such process. Cognitive dissonance occurs when there is a discrepancy or inconsistency between two attitudes, an attitude and a behavior, or between an attitude and a new piece of information. The discrepancy creates psychological anxiety or tension that we are motivated to alleviate. In order to restore psychological "balance," we have to change either our attitude, our behavior, or the new information. In other words, we persuade ourselves in order to bring our attitudes in line with other attitudes, our behavior, or with new information.

Social Influence

Getting others to behave in ways we want them to is a pervasive aspect of social relations and the subject of a great deal of research. *Compliance* is the name given to studies of the sorts of social influence where subtle, situational cues are arranged that increase the likelihood of agreeing to a request. People who have no particular authority over us are most likely to use these sorts of strategies to get us to comply with their requests (in other words, to get us to behave as they want). *Obedience* is the name given to studies where overt commands are used to generate behavior. People with authority over us have this option.

COMPLIANCE STRATEGIES

How do we bend others to our will? Psychologists have outlined several methods people use to gain compliance from others. You may recognize many of them because they are often used to separate us from our money or time.

Positive Mood

There are two compliance techniques based on eliciting a positive mood. Generally, it is easier to gain someone's compliance when they are in a good mood. The *luncheon technique* is based on the fact that it is easier to gain compliance from someone who is eating. People tend to be in a better mood while eating. It is no wonder that so much time and money is spent conducting business over a meal. *Ingratiation*, or saying something nice about a person, can also elicit compliance. Flattery works, even when the recipient knows it is merely flattery.

Reciprocity Norm

The reciprocity norm is a "rule" in our society that states that we return favors. If someone does something for us, then we feel compelled to do something for them. It is a very difficult rule to break, even if we never asked for the favor in the first place. There are three techniques based on the reciprocity norm.

The first technique is to merely exploit the norm by asking for a favor that is larger or more valuable than the favor you do for the other person. This is often used by charities that send you a "little something" such as address labels or holiday cards in the hope of receiving a donation. The Hare Krishnas were infamous for using this technique in airports in the 1960s and 1970s. They would approach a person, give them a flower, and ask for a donation. Most people tried to refuse the flower, but the Hare Krishna member would not take it back. Because we are taught to return a gift with a gift, many people felt obliged to comply with the request for a donation. Interestingly, many people threw away the flower after giving a donation, which was picked up by the member and given to someone else.

A second technique based on the reciprocity norm is known as the *door-in-the-face technique*. This involves first asking for a large unreasonable favor that will surely be denied, then requesting your smaller original favor. It takes its name from the inclination to slam the door in the face of the large request. However, when the large favor is requested first, a person is more likely to comply with the smaller favor than when the smaller favor is requested alone. For example, a teenager and her mother are shopping for the teen's first car. The girl sees the car she wants, but the price is a few thousand dollars over their budget. Instead of asking for the desired car, the girl finds a car that is $10,000 over their budget. Once the mother refuses the expensive car, the girl asks for the desired car, and the mother complies.

A third technique is known as the *that's-not-all technique*. This technique induces compliance by adding a benefit or dropping the cost while the customer considers the deal. This is often used in late night infomercials, from where it gets its name. It works by exploiting the fact that people feel pressure to reciprocate the deal because the salesman is taking a "loss" by offering something else of value. People are more likely to accept the sweetened deal than they are if it were the only offer. However, common business sense tells us that this added value is already factored into the cost of the item and is not actually costing the salesman anything.

Commitment

Compliance techniques based on commitment exploit the fact that we like to appear consistent to others. Thus, when we make a commitment to one request, we tend to commit to another similar request. This is particularly true if the commitment is made publicly. Two techniques are based on exploiting this tendency.

The *foot-in-the-door technique* relies on obtaining compliance to your original request by first asking for a smaller request. Once you get your foot in the door and get someone to comply with a small request, you can usually get them to agree to increasingly more demanding requests. For example, organizations that fundraise yearly often remind you of how much you donated last year and then ask for a slightly larger donation this year. Because you already publicly committed to a certain amount, you are more likely to increase your donation.

The *low-ball technique* is based on changing the cost of a behavior once you have already agreed to comply. Once a person has made a commitment to engage in a behavior, the cost of that behavior is raised or the value of the item is lowered. The person may still feel committed to the behavior and comply even though the cost is higher. For example, a car salesman may tell you that the price of the car you just agreed to purchase does not include a stereo, tires, or other extras. In order to receive those items, the price has to be raised. If the buyer has already committed to making the purchase with the extras, they will typically agree to the higher price even if they would not have agreed if that were the original price.

Reactance

Reactance occurs when one's freedom of choice is threatened. People feel an unpleasant state that motivates them to restore their freedom by performing the threatened behavior. Thus, when someone tells us that we cannot see, purchase, or obtain something, we tend to want that thing more than we did before that option was removed. Content media companies are notorious for exploiting reactance. Nothing sells faster than a censored CD, film, or book. A related trick used by salespeople is to suggest that the item for sale is out of the customer's price range. "You won't be able to afford this model," is just a way of telling us we cannot have it.

OBEDIENCE

Obedience differs from compliance in that the social pressures to obey an authority are stronger and may have more severe repercussions. The first and most influential experiments on obedience were conducted by Milgram.

Milgram conducted a series of famous experiments that were designed to determine whether the behavior seen during the Holocaust was due to the "evil" personality of the German people (a common notion among the public at the time) or, as he thought, the pressure to obey that was caused by the situation. To test this, he created a laboratory situation in which a subject would act as a "teacher" and a confederate (someone working with the experimenter pretending to be a research subject) would act as a "learner."

To study "the effect of punishment on learning," Milgram told the subject to give increasingly stronger doses of electrical shock for each mistake made by the learner. The experimenter was either one foot behind the subject or in the next room. The subject sat at a control panel with 30 different shock levels, with the highest level marked "XXX (450 volts - lethal)." The learner, in another room, made prearranged mistakes that necessitated shocks. At 300 volts, the learner began to pound on the walls and protest, as though he were in a great deal of pain. As the shock levels increased, the learner would cry for help, beg for the shock to end, and finally, stop responding altogether (as though he were unconscious).

If the subject turned to the experimenter for guidance or asked to stop the experiment, the experimenter responded authoritatively, "The experiment requires that you continue." Even though many subjects showed visible signs of discomfort, almost 65 percent of the subjects administered all 30 levels of shock (including the "XXX lethal" level) when the experimenter was next to them; a much lower percentage obeyed when the experimenter was in the other room.

Here are the factors that influence the tendency to obey:

- **Proximity to victims**   The farther you are from the victims, the more likely you are to obey an order to harm them.

- **Proximity to authority**   The farther the authority is from you, the less likely you are to obey an order to harm someone.

- **Institutional setting**   Taking the experiment out of the prestige of Yale University and into a run-down office building only reduces obedience slightly.

- **Conformity pressures**   Obedient peers can increase obedience, and rebellious peers can reduce obedience.

- **Gender**   Men and women are equally likely to obey an order to harm someone.

- **Cultural differences**   There are variations across cultures, but obedience rates are high across cultures.

- **Personality traits**   At best, personality traits are only weakly correlated with the tendency to obey.

Social Relations

Not only does the social context affect our behavior, but sometimes it defines it. Sometimes we behave only as a part of a group.

Next, we'll look at how people work in groups and the special case of that group of two that leads to sex and mating, *interpersonal attraction*.

WORKING IN GROUPS

The social psychology of groups is an important example of how basic science can work together with applied science. Basic research in social psychology on groups is used by applied psychologists in business and government to improve the effectiveness and efficiency of groups assigned to various tasks.

Tips on Terms: Group

A *group* is two or more persons working toward a common goal. That work may involve competition or cooperation between the group members, or both.

Jury Studies

What happens to peoples' attitudes when they discuss issues together in order to come to a decision? Juries are an interesting sort of group because they continue on until a consensus is reached. Early social psychologists actually studied real juries working on real court cases. (This would be ethically unacceptable today.) But they have also done many experiments where a group discusses a topic, with or without the requirement of coming to a consensus.

Typically, a topic is chosen and the subjects' attitudes on that topic are measured before and after the discussion to see how a particular initial configuration of attitudes affects the changes due to the discussion. A number of interesting phenomena have been found. For example, when all the members of the group have initially similar attitudes, their attitudes move together as a group. If all the members start with similar extreme attitudes, the attitudes become more extreme. If all the members start with similar moderate attitudes, the attitudes become more moderate. A group of extreme bigots becomes more bigoted. A group with more moderate attitudes on race becomes less bigoted.

Another effect is the *risky shift*. A confederate is placed in the group. The confederate espouses an extreme position and will not budge. Over time, the rest of the group's attitudes move toward the extreme presented by the confederate.

Groupthink

Historical studies are less common in social psychology than in other social sciences, but they have an important and growing role. An exemplary case is Janis' study of *groupthink*. Janis examined famously bad failures of groups working together and discovered commonalities that did not appear when groups worked effectively.

For example, the group that failed to protect the United States against the attack at Pearl Harbor and the group that decided on the disastrous military attack on the Bay of Pigs in Cuba exhibited the groupthink pattern of behavior. The attitudes of the individual members were similar initially and became more so. The group leader worked to narrow consideration to a small number of possibilities or alternatives whenever decisions had to be made. The group followed the leader closely. Group members chastised and teased individuals who did not follow closely. Individual initiative was replaced by a sort of collective will.

A large part of the groupthink mentality is a sort of misguided loyalty. Group members feel hesitant to "rock the boat" by expressing contrary opinions or supporting alternative courses of action. This hesitation to disagree leads group members to believe that they agree more than they really do. By way of contrast, in effective groups, the leader works to support members presenting alternative points of view and discussing the issues.

INTERPERSONAL ATTRACTION

Interpersonal attraction covers a myriad of relationships, from friendships to love. Each relationship is unique, but most relationships follow predictable patterns. Social psychologists studying relationships often use an economic model as a basis of theoretical analogy. They speak of the *rewards* and *costs* of relationships and how this influences *satisfaction* and *commitment*. This is not to say that they argue that interpersonal relationships are merely economic exchanges with "good feelings" as the currency, rather that they find the economic model useful for explaining how these factors influence the course of relationships.

Initial Attraction

The beginning of any relationship requires that two people meet. What factors contribute to initial attraction?

- **Proximity**   How physically close you are to another person. The closer a potential friend lives to you, the more likely you are to develop a friendship. You have to have contact with someone before you develop a relationship.

- **Similarity**   We tend to associate with people who are similar to us on several dimensions. These include social background, attitudes, values, and educational backgrounds.

- **Mere-exposure**   With all else being equal, people are attracted to those they've been exposed to more often. The more they've been exposed to a person, the more they tend to like them, but only up to a point.

- **Physical attractiveness**   This aspect rarely falls at the top of a list for potential friends or a potential date, but the effects are quite strong.

Fun Facts: Ducks and Swans

While most people tend to underemphasize the role of physical attractiveness in attraction, it has a strong effect on how we perceive the strength of relationships. We have a strong tendency to match people on the basis of physical attractiveness. Ducks belong with ducks and swans belong with swans. Further, the closer two people match in physical attractiveness, the more committed the relationship is perceived to be.

Building a Relationship

When a new relationship is developing, there are several factors that influence the level of commitment and the degree of satisfaction. Some theorists argue that this stage of a relationship is analogous to a social exchange of rewards and costs. Ideally, rewards should be high and costs should be low. However, the level of reward tends to be more influential than the level of costs incurred.

Continuation and Consolidation

This is the "middle age" or "settling down" stage of a relationship. The important factors determining commitment and satisfaction include the degree of communication, the level of communality in the relationship, and the amount of external support. During this stage, communication is important, particularly about the costs a partner is incurring. If one partner appears to be doing all of the "heavy lifting" in the relationship, satisfaction for that partner decreases. However, concern for the partner's rewards and costs, or the communality of the relationship, is better for the relationship. Finally, the external support from family, friends, and society is important for the relationship's continuation.

Deterioration and Decline

The deterioration and decline of a relationship is often determined by the available alternatives and the barriers to leaving.

First, a person will consider the attractiveness of the current relationship. Do the rewards outweigh the costs? If not, satisfaction with the relationship decreases and the likelihood of deterioration increases.

Second, do the alternatives to maintaining the relationship appear more attractive than the current relationship? These alternatives can include another partner, but often the alternative of being on one's own can be equally as attractive. If the alternatives appear more attractive than the current relationship, the likelihood of deterioration increases.

Third, people consider the barriers to leaving the relationship. Particularly in the case of divorce, the costs of leaving can be very high, both emotionally and economically. If the barriers are too high, the likelihood of ending the relationship decreases.

Finally, people consider the investment they have made in the relationship. When the proportion of "profits to investments" is unequal between partners, the relationship suffers.

Endings

The end of a relationship is rarely mutually desired. Typically, one partner decides to leave before the other. Are there differences between being the "dumper" and the "dumpee"? Yes. The person who decides to end the relationship has the advantage of time to prepare emotionally and practically for the split. They have had the time, while making the decision, to remove the barriers to leaving, find alternatives, and emotionally distance themselves from their partner.

Summary

Social psychology has focused its attention on measurable attitudes but has still managed to contribute a wide variety of extremely important results to both basic and applied psychology. This emphasizes the importance of the social environment to our behavior.

Quiz

1. If we attribute a cause of a behavior as being _____ and _____, we are likely to say that the behavior is due to ability.

    (a) Internal; unstable

    (b) Internal; stable

    (c) External; unstable

    (d) External; stable

2. The _____ is the general tendency to prefer internal attributions over external attributions.

    (a) Self-serving bias

    (b) Actor-observer bias

    (c) Fundamental attribution error

    (d) None of the above

3. Which of the following is not a message factor in the study of attitudes and attitude change?

    (a) Position advocated

    (b) Sidedness

(c) Repetition

    (d) Credibility

4. Which of the following is a compliance strategy based on mood?

    (a) Luncheon technique

    (b) Reciprocity

    (c) Door-in-the-face technique

    (d) Foot-in-the-door technique

5. _____ refers to the fact that people feel an unpleasant state that motivates them to restore their freedom by performing the threatened behavior.

    (a) Foot-in-the-door technique

    (b) Ingratiation

    (c) Reactance

    (d) That's-not-all technique

6. According to Milgram's studies of obedience, the farther an authority figure is from you, the _____ likely you are to obey an order to harm someone.

    (a) More

    (b) Less

    (c) Equally

7. Which of the following is an influential factor in initial attraction?

    (a) Mere-exposure

    (b) Physical attractiveness

    (c) Proximity

    (d) All of the above

8. When building a relationship, which is more influential?

    (a) Costs

    (b) Rewards

    (c) Attractiveness

    (d) None of the above

9. If the barriers to leaving a relationship are high, the likelihood of leaving
   _____.

   (a) Decreases

   (b) Increases

   (c) Stays the same

   (d) Is negotiated

10. Which partner has the advantage when ending a relationship?

    (a) The person being left

    (b) The person who decided to end the relationship

    (c) Both partners have equal advantage

    (d) Both partners have equal disadvantage

CHAPTER 3

Developmental Psychology

As people proceed through life, they undergo changes that are somewhat independent of experience. The developmental psychologist studies this process of change.

Overview

It is not just our bodies that change as we grow. Our skills and mental processes change as well. Developmental psychologists seek to describe and explain all the patterns of growth and change that occur in people from conception to death.

HISTORY

The empirical investigation into how children develop was begun by Tiedemann in the late 1700s. He authored a detailed biography of his infant child, recording the changes, both physical and psychological, during those first few years. Infant biography, an important first step toward developmental psychology, did not become popular until the latter part of the nineteenth century. By the dawn of the twentieth century, many such case studies were available.

The next step came when G. Stanley Hall (1844–1924), considered by many to be the founder of American psychology, began studying the recollections and reports of large numbers of children using questionnaires. Hall was influenced by the philosopher Rousseau. He believed that a child's psychological development echoed or *recapitulated* the historical development of human civilization. Just as society had presumably advanced from barbarism to literacy and commerce and criminal justice, so the child advanced from an uncontrollable little hellion to a polite and well-behaved member of the gentry.

CURRENT APPROACHES

In part because interest in research on children has come out of practical interests in children's health and welfare and how to raise them, developmental psychology has tended to stick closely to empirical evidence and has not gotten too involved in the theoretical fads of any particular period in psychology. Every culture has its own myths and rules about the best way to raise a child. An important goal of developmental psychology is to discover what really affects childhood development in order to structure child-rearing practices to ensure healthy and happy children growing into healthy and happy adults.

In no small part because we want to keep our children safe, and because human development is obviously different in important respects from the development of other animals, developmental psychology has always been more open to the methods of observation than other areas of psychology. This does not mean that developmental psychology is any less sophisticated or statistical or empirical. It just means that the experimental interventions so popular in the rest of psychology are often too risky to use on children. Despite this, long experience observing children has given developmental psychologists the expertise to be able to construct experiments with very young children, and even infants, as subjects. It is a mark of the maturity of psychology that it is able to study children safely and effectively.

Most recently, developmental psychology has come to recognize that developmental changes do not stop at the end of childhood. Our needs and our abilities

continue to change throughout our lives. Developmental psychologists now study the entire lifespan, still with an important focus on how this knowledge can be used to benefit each of us in each phase of life.

Cognitive Development

Cognitively speaking, tremendous development occurs during the first 12 years of life. By the time children enter school, they have mastered the intricacies of language, can count, recite the alphabet, and operate complex video games. By the time they leave the sixth grade, they are able to read, perform mathematical computations, and tell you about history and science. How do these changes—from the basic abilities present at birth to these intellectual abilities—occur?

BASIC ABILITIES

Compared to the abilities of other species in infancy, human infants are relatively incompetent. A human infant cannot walk, crawl, or explore its environment. They have very limited communication skills. They cannot protect themselves or feed themselves. However, as helpless as we are at birth, we are born with some basic abilities.

At birth, we have the beginnings of very complex sensory capabilities. Newborns are sensitive to the range of sound frequencies that include the female voice. They also have an acute sense of smell, and are able to coordinate information coming from different senses. For example, they can pair the sound of their mother's voice with the sight of her face.

Newborn infants are also born with some reflexes. These reflexes are automatic responses to certain stimuli and often aid in the infant's survival. For example, until about three months of age, putting a finger (or anything else) near an infant's mouth will trigger the sucking reflex. During this period, a light stroke on the cheek will trigger the rooting reflex, in which the infant will turn his head toward the stimulus and start sucking. Together, these reflexes help the infant eat.

Newborns also have reflexes that offer some protection from harm in the environment. Until approximately ten days of age, an infant will pull back from a sharp object. Until approximately five months of age, he will respond to a loud sound with a startle reflex. From birth and throughout life, an infant will reflexively blink when exposed to bright lights. Each of these reflexes helps to protect the infant.

Newborns are also born with rudimentary memory abilities. We have a tendency to assume that infants' memories are fleeting, but research shows that an infant's memory capabilities are better than we think. A two-month-old infant can maintain memories for a few days. By eight months, an infant can hold onto memories for approximately a week.

Given that humans are capable of retaining memories at such an early age, why is it that as adults, we cannot recall memories from our infancy? This phenomenon, known as *infantile amnesia*, is a type of blackout of episodic memory from the first few years of life. One explanation holds that we may have trouble retrieving memories because we lack the retrieval cues needed to recall these memories. The world of an infant looks very different from the world of an adult. A second explanation is that we may not be able to remember early experiences because infants may not be able to process information deeply enough. Specifically, infants do not process memories using language, which may be necessary for a memory to persist. A final explanation is that those early memories may be stored just as well as other memories, but are lost due to changes in the brain during development.

Finally, infants are born with an innate tendency to engage in a certain style of behavior. Their *temperament* may be the beginnings of personality. Infants tend to fall into one of four (or five, including the last listed below) temperament types:

- **Approach types** Those infants who generally approach new situations or stimuli and usually have a positive response to it.

- **Withdrawal types** Infants who typically react negatively to new situations, usually by crying or fussing.

- **Easy types** Infants who do not cry or fuss often and are not generally demanding.

- **Difficult types** Infants who cry often and are generally more demanding.

- **Slow-to-warm-up types** Infants who are restrained, unexpressive, or shy.

APPROACHES TO COGNITIVE DEVELOPMENT

A child's thought processes are different from an adult's thought processes in both quantity and quality. Thus, children think differently than adults in both the kind of thoughts they have and the complexity of thoughts they have. Two competing theoretical perspectives have developed to explain these differences: Piaget's view and the information-processing view.

Piaget's View

Beginning in the 1920s, Jean Piaget developed a very influential model of childhood development. Piaget viewed cognitive development as a series of qualitatively different stages. At each new stage, children construct a more mature view of reality, which, in turn, changes how they think about the world and assimilate new information. He argued that the child is essentially a different cognitive person during each of the four developmental stages.

Infants are born with *schemas,* or mental structures that organize perceptual input and connect it to the appropriate responses. As children mature, these schemas change in two ways. First, the schemas become more articulated in that they are more precise and complex. For example, the schema for sucking may contain more precise movements for locating the nipple of a bottle. Second, schemas become more differentiated in that a single schema can give rise to two different schemas. The schema for sucking may give rise to two different schemas, one for sucking a bottle and one for sucking a thumb.

Piaget also argued that how children move through the four stages of cognitive development depends on three interrelated processes: assimilation, accommodation, and equilibration.

Assimilation involves incorporating new information into an existing schema. For example, suppose a five-year-old thinks that "living things" refers only to animals, such as dogs, cats, birds, and people. When her parents take her to the zoo for the first time and tell her that an elephant is a living thing, she will be able to assimilate this information into her existing understanding of "living things." Thus, she may have to make only a minor adjustment for the size of "living things" in her schema.

Accommodation is the fundamental alteration of an existing schema to adapt to new information. For our five-year-old to understand that a tree is a living thing, she has to alter her concept of living things to be more abstract. She has to be able to include the qualities of growth and reproduction in her understanding of living things to make the new information fit.

Finally, *equilibration* occurs when the change in thinking allows the child to fit all of the pieces of her knowledge together. Her concept of living things incorporates every living thing she knows. Through the process of assimilation, accommodation, and equilibration, a child moves from one cognitive stage to the next.

Information-Processing View

The information-processing view focuses on the quantitative differences in how children process information. Thus, these theorists focus on how children mentally represent what they see and hear, how they operate on that information, and how their memories impose limits on the information they can handle. From this perspective, a ten-year-old's ability to understand concepts that elude a five-year-old is not due to a difference in how they construct reality. Instead, it is due to the older child's more advanced capacity for processing information. Cognitive abilities develop gradually, as opposed to Piaget's "giant step" manner.

COMPARING PERSPECTIVES

How does each theoretical perspective explain differences in each stage of development? In this section, we will examine each stage of cognitive development from both Piaget's view and the information-processing view.

Infancy

Piaget argued that the central cognitive task of infancy is constructing a view of the world that incorporates a basic understanding of objects and of cause-and-effect relationships. During the *sensorimotor stage*, children do not analyze problems and contemplate the consequences of their actions. Their understanding is derived solely from what they sense and do. Piaget further claimed that this stage progresses in a series of substages revolving around the concept of objects.

From birth to approximately four months of age, a child has no concept of the permanence of an object. When a toy that has captured a child's attention is partially hidden behind a piece of paper, the child will not search for the toy. They do not conceptualize the toy as existing when it is out of their view. It is not seen as an enduring object.

From four months to eight months of age, a child will begin to search for a partially hidden object, but not for a fully hidden object. They expect the partially hidden toy to exist in its entirety.

From eight months to twelve months, a child will search for a completely hidden object. However, they still have an incomplete concept of the object. If the toy is hidden behind two pieces of paper, the child will search behind only the first. At approximately eighteen to twenty-four months of age, a child has completely acquired object permanence.

The information-processing view argues that the behavior in these situations may reflect memory limitations and not the acquisition of object permanence. In order to perform Piaget's task, a child must hold the image of the toy in memory. In order to search for the toy, the child has to further coordinate that memory with both a visual and manual search. This coordination may be beyond the cognitive and physical capabilities of someone so young.

Preschool Age

Piaget argued that the cognitive task of the *preoperational stage* is for children to expand their world beyond the limits of their immediate perceptions. This stage is marked by representational thought, or the ability to represent things mentally when those objects are not physically present. Thus, a child learns to play "make believe" and imitate someone's behavior after that person has left the room.

This type of thinking also has its limitations. First, preschool-age children engage in *egocentrism*, which is the inability to take someone else's perspective. By perspective, Piaget meant the ability to literally see something from someone else's vantage point. Children at this age do not realize that others have a different perspective and that theirs is only one of many. They also engage in *animism*, which is the belief that all things are human, like themselves. Thus, they attribute human qualities to inanimate objects. For example, a child may think that the sun "sleeps" at night.

A third limitation is *centration*, or the tendency to focus on just one aspect of a problem while neglecting other important features. For example, preoperational children do not recognize that just because the shape of a ball of clay changes, its mass does not change. Centration has much to do with the inability to make conservations, which is important in the next stage of cognitive development. Finally, preschool-age children do not have the ability to reverse an action mentally. This is also important for the next stage of development. *Irreversibility* is the inability to mentally "undo" a change on an imaginary object. Using the clay example, a child would not be able to return the shape of the imaginary clay back to a ball.

The information-processing theorists argue that these patterns of thought are caused by the child's limited information-processing abilities. Children are able to understand the point of view of others, but only on a very simple level. Given the more complex tasks used by Piaget and his researchers, children are unable to perform.

Middle Childhood

Piaget argued that the major intellectual accomplishment during this period is the ability to perform concrete operations. These concrete operations are mental transformations on objects. This ability marks the loss of both centration and irreversibility from the previous stage. Children in the *concrete operational stage* gain the ability to perform mental transformations on an object, and then reverse it so that the mental object returns to its original state. The key to this stage is the recognition that certain features of things remain the same despite changes in other features. Piaget illustrated this ability by presenting children with various conservation tasks, some of which we outline here:

- **Conservation of number**   The understanding that even if some other factor of objects change, the number of objects in the set do not. Before children reach this stage, they are able to tell you that two equal parallel rows of pennies have the same number of pennies in them if both rows have the pennies touching each other. However, if one of the rows is elongated, the child will say that the elongated row has more pennies, even though each row still has the same number. They believe that number can vary with irrelevant changes in the length of the row.

- **Conservation of mass** The ability to understand that even though the shape of an object changes, its mass does not. Children begin to grasp the concept that two equal balls of clay have the same mass. When one is rolled out like a hot dog, it still has the same mass. Children who have not reached this stage would say that the mass is different.

- **Conservation of volume** The ability to understand that a liquid does not lose volume even though it is transferred to a different sized container. Children begin to realize that when one of two identical containers of liquid is poured into a taller thinner container, the volume remains equal. Children who have not progressed through this stage often say that the taller container contains more liquid.

- **Conservation of length** The ability to understand that the length of a string does not change even though its shape changes. When two equal strings are shown to a child and then one is bent, a child who has not progressed will report that the bent string is shorter.

The information-processing theorists argue that because many children can perform conservation tasks correctly at varied ages, Piaget may be off-base. Many children can perform number conservation at age six, but can't perform volume conservation until age eight. They reason that if the ability to understand conservation of number reflects the development of a new, more advanced logical structure, then why can't children immediately apply that new understanding to all of the conservation principles?

They further argue that the way children at this age use language may be part of the explanation. Children may assume that *more* is a synonym for *longer*, or that the experimenter's question about "which has more" is a question of length. The children may be interpreting the question as one that points out the salient or obvious change. Thus, they may possess the ability to perform conservation tasks at an early age, but they misunderstand the question.

Adolescence and Adulthood

Piaget argued that this stage is marked by the ability to perform mental processes using abstract thought and hypothetical situations. The *formal operations stage* is defined by the ability to use formal logic, deductive reasoning, and strategy planning. Adolescents are able to carry out systematic tests to solve problems. Consider the logic problem shown here:

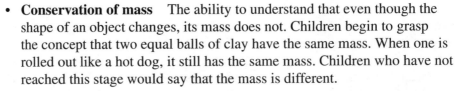

Premise 1	If Spot is a dog, then Spot walks on four legs.
Premise 2	Spot is not a dog.
Conclusion	Spot does not walk on four legs.

Is this correct? Can you deduce the conclusion from the premises? Piaget would argue that if a person has progressed through the formal operations stage, he will be able to tell you that the syllogism here is not valid. Spot may, indeed, be a cat.

The information-processing theorists point out a problem with this view. They note that even though people this age are able to carry out the logical operations necessary to solve and understand this type of problem, 75 percent of these people get the problem wrong. Further, large portions of the teen and adult population fail on other formal operational tasks. It might be that a person's performance on these tasks is due simply to having gradually acquired better information-processing skills. They are better able to use memory storage and retrieval cues and are better able to monitor their cognitions consistently.

Here is a summary of the Piagetan stages of development:

- **Sensorimotor stage** Found in infancy and characterized by the absence of analysis.

- **Preoperational stage** Found in preschoolers and characterized by the presence of representations.

- **Concrete operational stage** Found in school-age children and characterized by the ability to work with representations.

- **Formal operations stage** Found in adolescents and adults and characterized by the ability to do abstract reasoning.

Social Development

Beginning in the 1950s, a Freudian named Erik H. Erikson (he invented the last name himself) advanced a Freudian approach to development. Erikson viewed social development as a series of challenges, or *psychosocial dilemmas*, that people face at each of eight stages from infancy to old age. Each challenge has an outcome that is either favorable or unfavorable, which affects a person's social and personality development. A favorable outcome results in a positive outlook and positive feelings, which make it easier to cope with subsequent challenges. An unfavorable outcome leaves a person troubled and disadvantaged for future challenges. In addition to outlining Erikson's stages, we include a discussion of other factors affecting social development during each stage of the lifespan.

INFANCY

During the first year of life, Erikson argued that the challenge is to establish trust. In the *trust vs. mistrust* stage, an infant must depend completely on others to take care of the most basic needs, such as food, cleaning, and shelter. If an infant's basic biological and emotional needs are adequately met by her or his caregivers and secure attachments are formed, the infant should develop an optimistic, trusting attitude toward the world. If the infant's needs are not met, or are taken care of poorly, a more distrusting, insecure attitude may result.

During the second six months of life, infants begin to show signs of having developed enduring emotional bonds with their caregivers. This process may have a lot to do with the resolution of this stage of social development. Psychologists argue that there are four styles of *attachment*, defined as the emotional bond the child forms with her/his primary caregiver. Even though humans all seem to form attachments, these can vary in the quality of attachment. These variations are based on the type of care the infant receives. The attachment style is defined in terms of the way a child uses the primary caregiver as a base of exploration in her environment. The four styles are outlined here:

- **Secure attachment**   When the caregiver responds to the infant's needs promptly, appropriately, and consistently, the infant seems to acquire the expectation that the caregiver will be available and responsive. Securely attached infants use the caregiver as a base of exploration in that they are willing to venture away from them, while the caregiver remains in the room. They are upset when the caregiver leaves, are not easily comforted by strangers, but calm down quickly when the caregiver returns.

- **Resistant attachment**   When the caregiver is not consistent in responding to the infant's needs, the infant can develop a clinging unhealthy attachment. They do not use the caregiver as the base of exploration and instead tend to cling. They become angry when the caregiver leaves and are not easily calmed upon the caregiver's return.

- **Avoidant attachment**   When the caregiver is not consistent in appropriately responding to the infant's needs, the infant can develop an avoidant attachment to the caregiver. The infant seeks little interaction with the caregiver. They are equally comfortable with either the caregiver or a stranger.

- **Disorganized/disoriented attachment**   When the caregiver is not consistent in responding to the infant's needs, they can also develop a disorganized attachment. The infant becomes depressed and has periods of unresponsiveness along with spurts of sudden emotion.

EARLY CHILDHOOD

Erikson argued that from one year to three years of age, the challenge is to establish *autonomy vs. shame or doubt*. During this stage, parents begin efforts to regulate the child, such as toilet training. The child must begin to take some personal responsibility for feeding, dressing, and bathing. If the child masters this "beginner's independence," then the child acquires a sense of self-sufficiency. If the parents are never satisfied with the child's efforts and there are conflicts, the child may develop a sense of shame and self-doubt.

PRESCHOOL AGE

From the ages of three to six, Erikson argue that the challenge is for the child to establish *initiative vs. guilt*. In this stage, children begin to function socially within their family. They start to take initiative in trying out new activities. When this initiative brings them into conflict with others, guilt can result. Too much guilt can inhibit initiative, so children must learn to balance initiative against others' desires and needs. If the child learns to get along well with siblings and parents, a sense of self-confidence should develop.

MIDDLE CHILDHOOD

During middle childhood (from six to eleven years of age), Erikson believed that the challenge of social functioning extends beyond family to school and the neighborhood. During the *industry vs. inferiority* stage, children start learning the skills of their culture. Those who are able to function effectively in this less nurturing social arena, where productivity is highly valued, should develop a sense of competence. Those who fail may develop a sense of inferiority.

ADOLESCENCE

During adolescence, Erikson saw the main challenge as involving the struggle to form a clear sense of identity. The *identity vs. role confusion* stage is marked by the opportunity to develop an integrated sense of self as distinct from other people. Those who fail to do so tend to feel confused about their future roles in social interaction.

Much research has been conducted on adolescence since it was first identified as a stage of development. Some of that research has focused on how teens develop their self-identity. Identity formation involves the adoption of a sexual orientation, a set of ideals and values, and a vocational direction. A well-developed identity gives a person a sense of his strengths, weaknesses, and uniqueness. A person with a poorly developed identity is unable to define his personal strengths and weaknesses and does not have a well-articulated sense of self. Researchers claim that the presence or absence of crisis and commitment determine how teens cope with

Table 3-1   Identity Formation Orientations

		Personal Commitment	
		Present	**Absent**
Identity Crisis	**Present**	Identity achievement—successful formation of identity	Identity foreclosure—conforms to the expectations of society
	Absent	Identity moratorium—attempting identity formation, but requires more commitment	Identity diffusion—abandoned identity formation

identity formation. These are not stages, but are orientations that can occur. A person may go through any of these orientations or can become stuck in one phase. Table 3-1 illustrates these orientations.

EARLY ADULTHOOD

Erikson argued that the main concern of early adulthood is to develop the capacity to share intimacy with others. During the *intimacy vs. isolation* stage, young adults strive to form intimate friendships and romantic relationships with others. Those who succeed tend to be more empathetic and open to others. Those who do not succeed tend to feel lonely and isolated and are often manipulative with others.

This stage of adulthood is often when our internal social clocks begin to exert influence. Our social clocks are our individualized sense of personal development that specifies what we should have accomplished by a certain age. For example, some people want to be married by the time they are twenty-five. Others want to make their first million dollars by the time they are thirty. The accomplishments and their timeframe are largely personal, although culture often dictates when certain life-events should occur. For example, culture used to dictate that a woman who was not married by the time she was twenty years old was considered an "old maid" and was no longer a desirable candidate for marriage.

Big Background: Neural Pruning in Infancy and Adolescence

Until recently, researchers believed that the brain's anatomical development was complete at about one year of age. Now, using new, advanced brain-scanning technologies, researchers have discovered that the brain continues to develop at least until adulthood, with important changes occurring in late adolescence. Brain circuits that keep the adolescent from exhibiting self-control are *pruned* away, allowing the adult to better control his or her own behavior. Presumably, this is why young adults are less *impulsive* than adolescents.

MIDDLE ADULTHOOD

Erikson's challenge for middle adulthood revolves around developing *generativity vs. stagnation*. Generativity is the sense of responsibility to guide the next generation and be meaningfully productive in one's work. The focus is on raising healthy children, achieving a sense of satisfaction with vocational endeavors, and concern for what we leave behind. Stagnation occurs when self-indulgent concerns with meeting one's own needs and desires cause feelings of the meaninglessness of life. The person can become self-absorbed.

Given Erikson's description of this stage of life, it is not surprising that the midlife crisis typically occurs during this period as well. Traditionally defined, the midlife crisis occurs in men aged about forty to forty-five. It is the transition point in which one's perspective shifts from thinking about life in terms of how much time has passed between birth and now, to thinking about how much time remains. The crisis may lead one to end a marriage, change vocations, or make other major life changes. The midlife crisis is not, however, as universal as we may think. Not every man experiences a midlife crisis, and it may not be exclusive to men.

OLD AGE

Erikson's challenge during old age involves developing *self-integrity vs. despair*. Older people achieve a sense of self-integrity if they can reflect on their lives and see them as productive and satisfying. If, instead, they feel that their life was wasted and full of regret, they feel despair. This focus on the limited time one has left can greatly change their emotions and relationships. Older people tend to experience more positive emotions. They also tend to handle their emotions better than younger adults. Older people also tend to value emotionally satisfying relationships. They tend to limit their relationships to those with whom they are emotionally close.

DEATH AND DYING

While not really one of Erikson's stages of social and personality development, it is important to recognize this stage as distinct from old age. Psychologists examine death and dying from two perspectives.

First, we examine the response to the death of a loved one. Grief is the emotion of distress that follows the loss of a loved one. In most Western cultures, we typically experience three stages of grief. For approximately the first three weeks after the death of a loved one, the bereaved person is in a state of shock. They experience denial, disorientation, and eventually settle into a state of deep sorrow. From three weeks to a year, the person will experience a range of emotions, including anger, loneliness, and guilt. They tend to mentally review the relationship, wonder if they could have done something different, and look for the causes of their loss. Finally, by the beginning of the second year after their loss, the grief begins to lessen. The person may be able to start a new relationship, get through the day without thinking about their loss, and experience fewer and less severe symptoms of depression.

Second, psychologists examine the response to our own impending death. *Terror-management theory* states that death-related anxiety is the most fundamental source of personal anxiety. Most living things have an innate propensity to stay alive. Humans, of course, share that drive, but are the only creatures on the planet saddled with the knowledge that they will die. To buffer this anxiety, humans construct a shield. This shield is a belief system to explain how the world works. Living up to the values of our belief system bolsters our self-esteem. Thus, we manage the terror through a complex psychological defense comprised of faith, politics, and social norms that help us to feel safe.

In essence, we create and participate in culture in order to provide our lives with order, structure, meaning, and permanence. There are many methods in which culture gives our lives permanence and, therefore, provides us with a sense of immortality. We may attempt to secure eternal life by, for example, adopting a particular religious belief system. We may make contributions to society, be they ideas, tangible objects, or children. Through these methods, a person tries to ensure that they will live on in a symbolic sense.

Summary

Studying human development scientifically creates a challenge because development is governed by time. It happens no matter what the psychologist does. Thus, the role of the experimental intervention is limited. Despite this, developmental psychologists use longitudinal observation, theory, and experimentation to understand how our capacities for coping with the world change as we proceed through the stages of life.

Quiz

1. Which of the following is not a basic ability that human infants possess at birth?

    (a) Reflexes

    (b) Language

(c) Sensory capabilities

(d) Memory

2. Which of the following is not a possible explanation for infantile amnesia?

(a) The lack of retrieval cues

(b) Inability to process information deeply enough

(c) Memory lost due to brain development

(d) None of the above

3. _____ temperament types are those infants who generally approach new situations or stimuli and usually have a positive response to it.

(a) Approach

(b) Avoidant

(c) Withdrawal

(d) Easy

4. Piaget viewed cognitive development as a series of _____ different stages.

(a) Quantitative

(b) Qualitative

(c) Both a and b

(d) None of the above

5. _____ is the fundamental alteration of an existing schema to adapt to new information.

(a) Equilibration

(b) Assimilation

(c) Accommodation

(d) Articulation

6. According to Piaget, at which stage do children gain the ability to perform mental transformations on an object and then reverse it so the mental object returns to its original state?

(a) Sensorimotor

(b) Formal operational stage

(c) Preoperational stage

(d) Concrete operational stage

7. When the caregiver is consistent in responding to the infant's needs, the infant can develop a _____ attachment.

   (a) Secure

   (b) Resistant

   (c) Avoidant

   (d) Disorganized

8. According to Erikson, at which stage do children begin to function socially within their family?

   (a) Initiative vs. guilt

   (b) Autonomy vs. shame

   (c) Identity vs. role confusion

   (d) Intimacy vs. isolation

9. If an identity crisis is present and personal commitment is absent, which identity orientation is likely to result?

   (a) Diffusion

   (b) Moratorium

   (c) Achievement

   (d) Foreclosure

10. _____ theory states that death-related anxiety is the most fundamental source of personal anxiety.

   (a) Piaget's

   (b) Terror-management

   (c) Information-processing

   (d) Erikson's

CHAPTER 4

Intelligence

Psychologists develop a wide variety of tests to measure individual differences between people. Some of these tests measure things we can call *abilities*, in the sense that scoring higher on the test means that we are likely to do better at some task. A test of optimism/pessimism is not an abilities test. A test of memory is. There is one particular type of ability that is held in great regard, at least in Western society. It is called *intelligence*.

Overview

Intelligence is a hypothetical ability to manage well in school and in other activities calling for similar abilities. We all have a rough sense of what intelligence is, but establishing a scientific measure of intelligence is a very difficult and complex matter. Intelligence is not easy to define, particularly in terms of specifics. Generally, psychologists think of intelligence in terms of the ability to solve problems and to learn and understand complex material. Intelligence is more broadly thought of as the ability to adapt to our environment. In this chapter, we focus on how psychologists measure intelligence specifically and also explore how intelligence tests are developed.

HISTORY

Regardless of the contemporary definitions of intelligence, intelligence testing has been with us for a very long time. The first formal intelligence tests were administered in China approximately 4,000 years ago. Mandarin emperors screened candidates for government positions using a series of oral tests designed to find the most effective bureaucrats. Presumably, these tests captured some of what we would consider intelligence.

In the English-speaking world, the first major contribution to intelligence testing came from Sir Francis Galton. Galton's enthusiasm for assessing intelligence motivated him to collect various measures on more than 9,000 people at London's 1884 International Health Exposition. These measures included head size, visual acuity, color sense, and reaction time. Even though none of these measures is a valid indicator of intelligence, Galton should be recognized for his achievement. His contribution was not in his specific measurement techniques, but in the idea that individual differences in intelligence could be measured.

Modern intelligence testing began with Binet and Simon. In the early 1900s, the Paris schools were overcrowded. The Minister of Public Instruction decided to remove the slower learners from regular classrooms so that they could receive special instruction. However, at that time, only teachers could make that assessment, which was not reliable. The Minister appointed a commission to develop a test that could be administered to each child in the school system. Binet, with a history in assessment, was appointed to the commission.

Binet's approach was to give sample questions to large numbers of school children. If the correct answer to a question predicted good school performance in the following year, the question was used in the test. In this way, Binet constructed a test that effectively predicted future school performance, which was exactly what he was commissioned to do. The idea that Binet's test measured some general ability called "intelligence" was developed after the fact and by other people.

The test calculated an *intelligence quotient* (IQ score) in which a child's *mental age* (MA) was compared to her *chronological age* (CA). The equation was $IQ = MA - CA$. Children of normal intelligence would have IQ scores of 0 because their mental age would equal their chronological age ($0 = 5 - 5$). Binet originally defined "slow" as any child whose score on the test was two years or more below average for all children of that age. Thus, a seven-year-old who scored at a five-year-old's level would have met Binet's criteria ($-2 = 5 - 7$) for "slow," as would a twelve-year-old who scored at a ten-year-old's level ($-2 = 10 - 12$).

William Stern, a German psychologist, took the next step in the development of intelligence testing. He attempted to solve a problem with Binet's scoring of the test. The problem was that a twelve-year-old who was two years behind was considered just as "slow" as a six-year-old who was two years behind. However, intuitively, the six-year-old might seem more "slow" because the two-year disparity is a greater proportion of the child's total age. Stern believed that, instead of using the absolute difference between mental age and chronological age, testers should use the ratio between the two (and then multiply by 100 to avoid fractional scores). His new equation was $IQ = (MA/CA) \times 100$. Thus, the difference between the scores of the twelve-year-old ($83.3 = (10/12) \times 100$) and the six-year old ($66.7 = (4/6) \times 100$) can account for the difference in the proportion of age.

Binet's scales and Stern's ratio were greeted with more enthusiasm in the United States than they were in France. Many states had passed laws that required children of "normal" intelligence to attend school until the ages of 13 or 14. Intelligence scales rapidly became enormously popular in the United States, because they now had some measure for identifying these children. However, these scales had to be standardized and normalized for US children. This job fell to Stanford University psychologist Lewis Terman, who developed the Stanford-Binet intelligence test.

By the early 1900s, the United States entry into WWI brought about the next major development in US intelligence testing. The military requested a method of classifying its 1.5 million recruits in order to determine who should receive officer training, regular training, or be rejected. The Army psychologists revised the Stanford-Binet test (among some lesser-known others) so that it could be given to large groups simultaneously. This began the US fascination with large-scale, institutionalized testing.

The final major development in the history of intelligence testing, particularly in the United States, was due to David Wechsler, a psychologist in the 1930s. Because of his position as chief psychologist at New York City's Bellevue Hospital, he needed to assess the intelligence of his adult patients. The Stanford-Binet test was designed to assess the intelligence of school-age children. It was not valid for adults. He developed the Wechsler Adult Intelligence Scale (WAIS). The WAIS tested a wider variety of abilities and expanded intelligence scoring to include adults. It is important to note that the area of intelligence testing continues to advance, with new theoretical perspectives on intelligence and new tests being developed in line with those perspectives.

With the establishment of effective and reliable means of measuring intelligence came research into the nature of intelligence. Early on, researchers used sophisticated statistical measures to see if they could determine whether or not there was a single underlying factor behind all of the skills that led to success in school or in life. Spearman argued that intelligence was a reflection of one general factor he called "*G*." Thurstone argued that there were multiple factors underlying intelligence.

CURRENT APPROACHES

More recent research has focused on the degree to which intelligence is based on genetic or environmental causes. There have been attempts to predict intelligence in terms of highly complex reflexes. Most recently, the notion that there are multiple kinds of intelligence, not just school-book intelligence, has come to the forefront of research. While this research is reminiscent of the earlier debates over *G,* that earlier research was about various skills useful in an academic setting. The newer notions of multiple intelligence are about skills not traditionally characterized as "intelligence."

Test Construction and Evaluation

Constructing a psychological test is not as easy as collecting a set of relevant questions and giving them to other people. There are special considerations that psychologists have to follow in order to develop a good test. A good test is presented the same way to everyone who takes it, consistently gives the same scores for an individual, and measures what it is supposed to measure. In other words, a good test is *standardized, reliable*, and *valid*.

STANDARDIZATION

The process of standardization involves developing uniform procedures for giving and scoring a test. Ideally, everyone who takes the test is subject to identical conditions. They receive the same instructions, the same amount of time to take the test, the same materials, and the like. To begin the process of standardization, a psychologist will first administer the test to a large number of people under uniform conditions. The second step is to determine the *norms*, or the established standards of performance on the test. This can be done by administering the test to a large group of people who are representative of those for whom the test is intended.

Once the test is scored, the mathematical distribution of the scores is determined. This allows the psychologist to determine whether or not a score is meaningful. For example, if you took a test and scored a 29 without knowing the norms, you would not know if that was a good score or a bad score. Through standardization and norming, the average can be calculated, and we can know whether the score of 29 is average, above average, or below average. Standardization and norming give meaning to an individual's score.

RELIABILITY

Reliability refers to the consistency of a person's scores. Standardized psychological tests generally assess factors that are not expected to change over time, such as personality traits and intelligence. The consistency of an individual's score is important for the meaningfulness of the test. If a person scores 15, then 119, then 78 on three different occasions, then something is likely wrong with the test. Consistency, or score stability, is established by retesting the test taker with the same test on two different occasions. A correlation coefficient is calculated, and the closer the correlation is to +1.00, the more reliable the test.

This *test-retest* reliability is the most basic form of reliability testing. The identical test is administered on two occasions, usually a day to several weeks apart. The test-retest reliability is high if the score on the second test is similar to the score on the first test. The problem with this form of reliability testing is found in interpreting the second score. It is likely that the person recognizes the questions and remembers how they answered the first time. Thus, the test may be saying more about the memory abilities of the test taker than about the reliability of the test.

To avoid this problem, psychologists use variants of test-retest reliability testing. *Split-half* reliability testing involves splitting the test into two halves and administering one half of the test at each testing session. It is rarely the case that a psychologist will devise a test in which there is only one question for the target evaluation. Typically, there are 10–20 questions designed to evaluate a specific factor, and possibly more depending on the size of the test. The questions for each factor being evaluated are sorted, halved, and used to create two smaller tests. The scores for each factor are then averaged and correlated with the other half of the test.

A second variation is to create different versions of the test. *Alternate-form* reliability testing is a little trickier. It is difficult to write two questions that are different enough so as not to trigger memory, but are both tapping the same concept. Once this problem is solved, the psychologist will give one version of the test at the first testing session, the second version at a second testing session, and then correlate the scores.

VALIDITY

Even though a test is standardized and reliable, it still has to measure what it is supposed to measure. For example, an intelligence test based on measuring skull size could easily be standardized and tested for reliability. The test could be administered to a large group of people and averages calculated. The reliability of the test could be determined by measuring a person's head many times. The correlation coefficient would be very high, indicating high reliability. Thus, this intelligence test is standardized and highly reliable. However, is it really measuring intelligence? No. Head size is not a *valid* measure of intelligence.

There are several types of validity that can be measured:

- *Criterion validity* is important for tests that are attempting to identify people who have a specific trait or ability. A person's score on a test is correlated with some other yardstick of the factor that the test presumably measures. The yardstick, or criterion, is the measure against which the test scores are compared. For example, you could criterion-validate the head-size intelligence test by comparing your results with other valid intelligence tests.

- *Content validity* involves determining the test's ability to cover the complete range of material that it is supposed to cover. Any test that covers material it is not supposed to cover, or any test that doesn't cover everything it is supposed to cover, would not pass this validation step. For example, you could content-validate the head-size intelligence test by ensuring that it measures each factor of your definition of intelligence.

- *Predictive validity* involves determining the test's ability to predict future performance. The SAT, ACT, LSAT, GRE, and MCAT are all attempts to predict a person's future educational performance.

It is important to note that not all types of psychological testing require each type of validity. It is up to the test's author to determine the types of validity that are relevant and then to assess those types.

Assessing Intelligence

Currently, the two most common intelligence tests are the Stanford-Binet intelligence test and the Wechsler Scales. There are many other types of intelligence tests, but these two are the tests that most people will encounter during their life.

STANFORD-BINET INTELLIGENCE TEST

The Stanford-Binet intelligence test is the updated version of the original Stanford-Binet test discussed previously. This test is designed to assess the intelligence of children and adults, ages two to twenty-four. The items on the test range in difficulty and are arranged by age. Thus, there are two-year-old's questions, five-year-old's questions, sixteen-year-old's questions, and so on. These questions are defined as questions that can be answered by 60–90 percent of children of that age.

The test is administered to an individual in a standardized manner. The tester begins by asking the individual the questions for their chronological age group. If the individual successfully answers most of the questions, the tester moves up to the next higher age group's questions. This continues until the individual begins to miss questions regularly. If the individual starts out by answering questions incorrectly, the tester moves down to the next lower age group's questions. When the individual begins to repeatedly miss item after item, the test is terminated.

Performance on the Stanford-Binet is scored and transformed into an IQ score. That score is standardized so the mean equals 100 points and the standard deviation equals 15 points on a normal distribution (refer to the Chapter 1 for more information). From this, you can plug the individual's mental age score into Stern's equation to calculate the IQ score. For example, a ten-year-old who scores at a ten-year-old's level would have an IQ score of 100 (100 = (10/10) × 100). This score would be average, indicating that this child is performing at the level appropriate for his age. A ten-year-old who scores at a fourteen-year-old's level would have an IQ of 140 (140 = (14/10) × 100), which is above average. A 10-year-old who scores at a six-year-old's level would have an IQ score of 60 (60 = (6/10) × 100), which is below average.

WECHSLER SCALES

The Wechsler Scales (WAIS and its derivatives) are one of the most frequently used individual intelligence tests and are among the most valid and reliable. They include several different tests and which one is used depends on the age of the person being tested. The three main Wechsler Scales are as follows (the *R* indicates a revision):

- **Wechsler Adult Intelligence Scale (WAIS)**   The intelligence test for adults.
- **Wechsler Intelligence Scale for Children (WISC-R)**   The intelligence test for ages six through sixteen.
- **Wechsler Preschool and Primary Scale of Intelligence (WPPSI-R)**   The intelligence test for ages 4 to 6.

Each Wechsler Scale includes two subtests. These subtests allow the psychologist to derive an IQ score for each category, as well as an overall IQ score. The *performance* subset tests abilities such as picture arrangement, picture completion, object assembly, and matching digits with the symbols with which they were previously matched. The *verbal* subset tests abilities to answer general information questions, general comprehension, arithmetic, and vocabulary.

Using Stern's ratio is problematic with the Wechsler Scales, particularly the WAIS. If, as is assumed by all intelligence tests, intelligence is stable over time,

Table 4-1   The Problem of Mental Age for Adults

Chronological Age	Mental Age	Stern's Ratio	IQ
20	20	$(20/20) \times 100$	100
25	20	$(20/25) \times 100$	80
30	20	$(20/30) \times 100$	67
35	20	$(20/35) \times 100$	57
40	20	$(20/40) \times 100$	50
45	20	$(20/45) \times 100$	44
50	20	$(20/50) \times 100$	40
55	20	$(20/55) \times 100$	36
60	20	$(20/60) \times 100$	33
65	20	$(20/65) \times 100$	31

then Table 4-1 demonstrates the problem. If an individual is tested every five years throughout their adult life, and their score remains the same, their intelligence appears to decrease incrementally. Commonsense tells us that this is not the case. Stern's ratio is mostly a problem with adult measures of intelligence because it is very difficult to distinguish between abilities one should possess at forty years of age versus fifty-five years of age. Stern's ratio is based on the rapidly increasing skills found in school children. It simply doesn't make much sense for adults.

Therefore, scoring the Wechsler Scales is no longer based on Stern's ratio. The new scoring methods rely on how an individual scores relative to the performance of others of the same age. To maintain consistency between the Stanford-Binet and the general common notion of what an IQ score means, 100 is still average.

Influences on Intelligence

Why are there individual differences in intelligence? What makes some people smarter than others? Researchers have investigated various biological, genetic, and environmental explanations for these individual differences.

THE BRAIN

Psychologists have found that the relationship between the brain and intelligence is a complex one. The two most promising avenues of research involve brain size and brain speed.

Recent studies have demonstrated that the larger a person's brain, then the higher their IQ score tends to be. Does this mean that the head-size intelligence test mentioned earlier might be valid after all? No. First, head size is only a relatively accurate indicator of brain size. Second, even brain size correlates with intelligence only to a certain degree. Third, it is actually the ratio of brain size to body weight that gives the highest correlation to intelligence.

Finally, we simply cannot claim that a larger brain causes a higher IQ score. We cannot determine whether larger brains cause us to act more intelligently or if acting intelligently causes our brains to be larger. The size of crucial areas of the brain that control certain behaviors measured by intelligence tests may be more important than overall size. If a person scores high on the verbal section of an IQ test, they might have more neurons in the area that controls verbal comprehension. However, studies have also demonstrated that males and females have the same relative intelligence scores, but female brains are generally smaller. Thus, this remains an open question.

Brain speed has offered a promising avenue of study for intelligence researchers. Studies have shown that people with higher IQs tend to process information faster than do people with lower IQs. Again, we cannot draw a direct causal inference from this relationship alone. It might be that people with high IQs process information faster at certain steps in the problem-solving process, but are not faster overall. For example, researchers have found that people with higher IQs actually tend to spend more time digesting a problem and figuring out what kind of reasoning will be needed. Thus, it may be the efficiency of information processing and not merely the speed that makes the difference in intelligence.

GENETICS

It has long been argued that there is an important genetic contribution to intelligence. Thanks to new technology and research related to the Human Genome Project, new methods may soon add light to this old controversy.

How do we tease apart the relative contribution of genes and other factors? Typically, researchers use various types of heredity methodologies. Adoption studies compare the IQ scores of adopted children to their adopted relatives and to their biological relatives. The best twin studies for intelligence testing are to study identical twins who have been separated at birth. This helps to eliminate many environmental confounds that come with being raised in the same household.

Based on these methodologies, several lines of research support the notion that there is a genetic component to intelligence. First, the correlations of IQs of adult identical twins raised apart are higher than the correlations between fraternal twins or nontwin siblings raised together. Second, the IQs of adopted children correlate higher with their biological mother's IQ score than they do with their adoptive mother's IQ. Finally, by the time adopted children and their adoptive siblings are adults, there is no significant correlation between their IQ scores. These results indicate that there is a genetic component to intelligence.

The new technologies of the Human Genome Project should allow psychologists to identify individual genes that contribute to intelligence. Although this research is in its infancy, we already know that there is no single gene, or even a small set of genes, that determine intelligence. Research to date has not found good correlations between identifiable DNA patterns and intelligence. There could be a number of different reasons for this. It may be that multiple genes contribute to intelligence in different ways. Or it may be that the genetic contribution to intelligence is deeply interwoven with the environmental contribution. If certain genes make us smarter only in certain environments, then the mathematical assumptions underlying the statistics of heredity studies, including twin studies, do not hold. In that case, the actual amount that genes contribute to intelligence is unknown.

Critical Caution: Heritability

Heritability does not mean what you think it means.

Twin studies, and other studies that try to determine the influence of genes on behavior without actually examining the genes, use a statistical measure, called *heritability*, which is usually interpreted as the proportion of something that is due to genetic influence. But what does it really mean to say that 80 percent of intelligence is due to genetic influence?

Consider the following: The heritability index for anorexia nervosa and bulimia (two serious eating disorders in which sufferers either starve themselves or purge themselves of food) is reported to be between 50 percent and 80 percent, with a recent study placing it at 57 percent. Does this mean that 57 percent of people with these disorders have them due to genetics? No. Does it mean that, on average, a person with one of these disorders is 57 percent more likely to have developed it due to genetic factors? No. Does it mean that, on average, no more than 43 percent of the severity of the disorder can be altered by environment? No.

Unlike schizophrenia, depression, mental retardation, and many other disorders, anorexia and bulimia were almost unknown until modern times. The first reported case of either dates back to the thirteenth century. (There are mentions of unusual individuals who may have had similar traits in Ancient Greece and Rome.) It was not recognized as even a rare disorder until the seventeenth century. It did not become common enough to attract clinical attention until the 1960s and 1970s and still may be extremely uncommon outside of the industrialized world. There is no possibility that the genes in the human population have changed that fast. If a person was born 400 years ago with the exact same DNA as someone today who has severe anorexia or bulimia, there is almost zero probability that she would have had the disease.

What else besides genes could affect the heritability index? Both anorexia and bulimia are correlated with high self-reports of childhood sexual abuse. Assume, for a moment, that sexual abuse can increase the likelihood of anorexia and bulimia (although this is far from being proven). A sexual predator with access to a child is more likely to have access to his or her sibling. Sexual predation runs in families because being sexually abused as a child increases the likelihood that one will abuse children as an adult. If a sexual predator has any preference based on looks, then he is more likely to abuse both of a pair of *monozygotic* (identical) than both of a pair of *dizygotic* (nonidentical) twins.

People born in the same place and the same time tend to be treated similarly. People who look more alike (as monozygotic twins do) are treated more similarly than those who look different. Twins are treated differently than other people. All of these things tend to increase the heritability index. Very few tend to decrease it.

But the one thing that can *most* affect the heritability index is the interaction between genetics and environment. Complex interactions are not properly measured by the heritability index and can cause it to rise or fall dramatically. It might be that there is a gene that causes anorexia or bulimia, but that it has its effect only in specific social environments, and none of those environments existed on Earth until modern times. The precise number of 57 percent might not mean much at all.

ENVIRONMENT

Even though intelligence has a genetic component, this does not mean that the environment does not also play a large role. Genetics, brain biology, and the environment all contribute to intelligence. Because of the nature of heredity studies, it is difficult to determine how much is genetics and how much is environment or how much they interact.

For example, identical twins separated at birth may still share some of the same cultural norms. They are likely to share a similar school experience, similar cultural experiences, and similar social experiences. These similarities artificially inflate the estimate of the genetic contribution to intelligence found in twin studies.

Additionally, siblings raised together have one very important environmental difference: the other sibling. Think of a home with two children. The elder child has no older sibling and the younger child has no younger sibling. Only the house and the parents and the pets and so on, are identical. This difference artificially lessens the estimate of the environmental contribution to intelligence.

Thus, heredity studies do not conclusively distinguish genetic and environmental influences on intelligence. We must await the new DNA-based research.

Fun Fact: The Flynn Effect

The *Flynn Effect* is that population IQ scores have been increasing over the decades. Flynn found that in the Western world, IQ scores have generally risen about 3 points every 10 years. If we measured IQ the same way we did 50 years ago (without adjusting the average back to 100), the average IQ score would be 115. Why? Researchers offer three possible explanations. First, daily life is more complex, and the act of coping with these complexities may have increased IQ scores. We multitask more now than people did 50 years ago. Second, nutrition is better, and the improvements that come with improved nutrition may have increased IQ. For example, the average height of people has increased along with IQ scores. It may be that improvements in brain functioning may have also improved. Finally, the types of reasoning that IQ tests measure have improved, resulting in better test performance. For example, the use of technology in schools, consumption of mass media, and multitasking may have increased our ability to think abstractly and to think more quickly.

Types of Intelligence

Thus far, we have discussed intelligence as though it were a single thing. Several theorists have proposed that there are several types of intelligence, each of which can be measured. We will discuss the contributions of Sternberg's *Triarchic* theory of intelligence and Gardner's theory of *Multiple Intelligences*.

TRIARCHIC THEORY OF INTELLIGENCE

Sternberg has developed one of the more comprehensive theories in the field of intelligence. Traditional research on intelligence focused on individual differences in test scores. This approach to intelligence emphasized products or the end results of intellectual work. In contrast, Sternberg focused more on the process of intellectual work by emphasizing an information-processing approach. The Triarchic theory specifies three important components of intelligence. A person can excel in one, two, or all three types of intelligence.

Analytic intelligence involves the mental processes used in thinking. This corresponds most closely with the traditional notions of intelligence as measured by intelligence tests. Analytic intelligence is made up of three components. First, the *metacomponents* are the higher-order processes that define, plan, monitor, and evaluate a problem. For example, in the intellectual task of writing a paper, you plan your general strategy for the paper, you monitor your writing progress, and you evaluate how well the finished paper meets its goals. Second, the *performance*

components are lower-order processes that actually carry out the commands of the metacomponents. For example, the performance components of writing a paper would include assembling the words and sentences on the pages of the paper. Third are the *knowledge-acquisition components*, which are the processes involved in learning and storing information. These correspond to researching the paper.

Sternberg also proposed *creative intelligence*, which focuses on how people perform tasks with which they have either little or no previous experience. One student may be able to understand and comment on the theories he is taught in class. He may be able to explain them well to other students. But he may not generate many new ideas of his own, even in areas in which he is highly expert. Another student may not be able to grasp the theories found in the book, but may be able to come up with sound, novel ideas for theories of her own. This is creative intelligence, the ability to generate novelty.

Sternberg associates these creative skills with the ability to handle novelty and the speed with which a person automatizes routine tasks. A person who is high in creative intelligence can solve a novel problem easily and can figure out how familiar tasks can be performed more automatically. For example, a person may have average grades in college, but she is able to figure out how to use the U-Bahn in Germany without speaking any German. She may also quickly learn to drive without "thinking," leaving extra cognitive capacities available for other activities, such as listening to music or talking to the passenger.

The final type of intelligence is *practical intelligence*. This type of intelligence involves what is commonly known as "street smarts." More formally, it is the ability to adapt to, select, and shape one's real-world environment. Flexibility in changing one's behavior to cope with new situations, ease in taking initiative to change the environment to better suit one's behavior, and the ability to adjust to wholly new environments are the hallmark of Sternberg's view of practical intelligence.

THEORY OF MULTIPLE INTELLIGENCES

Gardner's theory of Multiple Intelligences posits that there are eight types of intelligence that involve a collection of abilities working together. Strength in a type or types of intelligence allows us to solve problems and to understand complex material within that type. We can be strong or weak in several types of intelligence. The eight types of intelligence are outlined here:

- **Linguistic intelligence**   The ability to use written or spoken language well
- **Spatial intelligence**   The ability to reason well about spatial relations
- **Musical intelligence**   The ability to compose and understand music
- **Logical/mathematical intelligence**   The ability to manipulate abstract symbols

- **Bodily/kinesthetic intelligence** The ability to plan and understand complex sequences of movements
- **Intrapersonal intelligence** The ability to understand one's self
- **Interpersonal intelligence** The ability to understand other people and social interactions
- **Naturalistic intelligence** The ability to observe carefully and interact with patterns in nature

This theory has a mass appeal in popular culture. It is commonly used in business situations to match an employee to a type of task or to match people with jobs. Gardner himself has said that he never intended for his theory to be used to profile people. Many of these intelligence types cannot yet be reliably measured and may reflect skills and talents rather than intelligence. However, the theory does provide a novel way to organize how we think about the study of intelligence.

EXTREMES OF INTELLIGENCE

The extremes of intelligence, retardation and giftedness, include people who score 2 standard deviations below or above the mean on an IQ test. Thus, they either have an IQ score of 130 or above in the case of giftedness, or 70 and below in the case of retardation. Because intelligence is normally distributed, this includes 4.56 percent of the population. We have much more information about retardation because the original intent of intelligence testing was to determine which school children needed to be removed from the classroom and given special help. Giftedness is a relatively neglected area of study, with only a handful of school programs available for those who test as gifted.

Variant View

Traditional psychometricians have been skeptical of both Sternberg's and Gardner's cognitive approaches to intelligence. According to the traditional view, irrespective of how the different types of intelligence work, it is critical to be able to measure each type of intelligence and demonstrate statistically that it is separate from the others. Despite the fact that Gardner's theory is twenty-five years old, his well-funded institute has not produced tests that measure his eight different types of intelligence, much less that demonstrate that they really are different. Sternberg has been having more success. After ten years, a test of Triarchic intelligence called STAT is reportedly doing better than the SAT for predicting college performance. Of course, neither the test nor the study is yet published, so we will have to wait and see.

Retardation

Mental retardation cannot be classified by IQ score alone. A person must meet three requirements before he receives the classification. First, intellectual functioning must be significantly below average, which is an IQ score of 70 or below. Second, the person must have difficulty functioning in at least two everyday activities, such as caring for himself or communicating with others. Finally, the onset of these difficulties stems from childhood. All of these requirements must be present in order for someone to receive a mental retardation classification. Thus, if a child with a 60 IQ could dress herself and show fairly normal motor skills, she would not be classified as mentally retarded.

There are four general categories of mental retardation. These categories demonstrate that there is a great deal of diversity and variation within mental retardation. Those categorized with mild retardation typically fall within the 50 to 70 IQ range. They are generally able to complete sixth-grade academic work and can hold a job in a supportive setting. Those categorized with moderate retardation typically fall within the 35 to 50 IQ range. They are generally able to complete second-grade academic work and can hold a job in a structured workshop-type setting. Those categorized with severe mental retardation typically fall within the 20 to 35 IQ range. Generally, they can learn to talk and often need help and supervision for even simple tasks. Finally, those categorized with profound mental retardation typically have an IQ score of 20 or lower. They generally have little or no speech abilities and require constant care.

The causes of mental retardation are varied. There are over 350 known causes, yet we can categorize them into two distinct groups: *Organic* retardation is caused by a genetic disorder or some physical damage to the brain. This damage can often be traced to an infectious disease or medical complications from a premature delivery. Down Syndrome is the most common form of genetic mental retardation. It is caused by the presence of an extra chromosome. This genetic abnormality prevents the neurons from firing normally. The degree of mental retardation varies widely, and may disrupt normal activities less than other forms of retardation.

Cultural-familial retardation can be traced to environmental factors, rather than genetic or biological causes. People with this type of mental retardation usually come from environments that are psychologically, socially, and economically impoverished. In general, their housing, nutrition, and medical care are also inadequate. Fetal alcohol syndrome is a common example.

Giftedness

There is much psychologists do not know about giftedness. Although there is no firm cutoff for IQ score guidelines, most psychologists categorize giftedness beginning at 130. However, most of the research has focused on the high end of the scale, such as 150 or above. Generally, people who are classified as gifted have an above average ability in some area. Prodigies, for example, are very talented in one area (music or mathematics) but are normal in others.

People classified as gifted may also be above average in creativity. They are able to intuit solutions to problems quicker and more effectively than people of normal intelligence. They initiate and pursue original ideas rather than staying on the assigned task. They also tend to have a greater task commitment in that they have a highly focused motivation when completing a task. Those who are gifted are particularly good at planning, organizing, and transferring strategies from one task to another. They tend to process information more effectively. Unfortunately, they also tend to be more prone to social and emotional problems due to mistreatment by their peers and parental expectations.

In the 1930s and again in the 1960s, there have been movements to better understand and assist gifted children. The idea was that helping the gifted attain their full potential would benefit all of society when those gifted children reached adulthood. Since the 1960s, there has been a pervasive concern for lifting the less able up to the status of those of average ability and very little concern about getting the best from the most able.

Summary

Intelligence, as a concept, has been a moving target since it first became a subject of psychological study over a hundred years ago. Despite its origins in the practical concerns of school enrollment and the enormous sophistication of the statistical methods used in testing, it remains a controversial area. Perhaps this is because the people who study intelligence make their livings being smart. Everyone wants a theory of intelligence that defines themselves as intelligent. Psychologists who theorize about intelligence are in a position to make that happen for themselves.

Quiz

1. Binet originally defined "slow" as any child whose score on the test was _____ or more below average for all children of that age.

   (a) One year

   (b) Two years

   (c) Five years

   (d) Seven years

2. In the equation IQ = (MA/CA) × 100, MA refers to _____.

    (a) Mental aggregate

    (b) Mental acuity

    (c) Mental age

    (d) All of the above

3. Establishing _____ of a test can be done by administering the test to a large group of people who are representative of those for whom the test is intended.

    (a) Reliability

    (b) Validity

    (c) Norms

    (d) None of the above

4. _____ refers to the consistency of a person's scores.

    (a) Reliability

    (b) Validity

    (c) Norms

    (d) All of the above

5. _____ reliability occurs when different versions of a test are administered on two occasions.

    (a) Split-half

    (b) Test-retest

    (c) Criterion

    (d) Alternate-form

6. _____ validity is tested by correlating a person's score on a test with some other yardstick of the factor that the test presumably measures.

    (a) Predictive

    (b) Content

    (c) Test-retest

    (d) Criterion

7. Which of the following are not subsets of the WAIS test?

    (a) Math

    (b) Verbal

(c) Performance

(d) None of the above

8. Which of the following influences our level of intelligence?

(a) Brain size

(b) Brain speed

(c) Genetics

(d) All of the above

9. According to Sternberg's Triarchic theory of intelligence, which of the following is not a type of intelligence?

(a) Analytic intelligence

(b) Creative intelligence

(c) Linguistic intelligence

(d) Practical intelligence

10. According to Gardner's theory of Multiple Intelligences, the ability to understand other people and social interactions is _____ intelligence.

(a) Naturalistic

(b) Interpersonal

(c) Intrapersonal

(d) Spatial

CHAPTER 5

Personality

People differ over and above how they look, their physique, their talents and skills, their social status, how well they do in school, and so on. People also differ in their character. Personality psychologists study the differences in character and the causes of those differences.

Overview

People often act as amateur psychologists when they meet someone new. We try to assess the new person's personality because we like to be able to predict how that person will behave in future interactions. We often describe others and ourselves in terms of personal characteristics, but not all of these characteristics are a part of our personality. What is personality? First, *personality* is the tendency to behave, think, and feel consistently over time and across situations. If introversion is a part of your personality, then you will tend to be introverted over a variety of situations and over much of your lifetime. Second, our unique collection of personality characteristics sets us apart from other individuals. Therefore, a person's personality consists of all the relatively stable and distinctive styles of thought, behavior, and emotional responses that characterize a person's adaptations to surrounding situations. In other words, personality is a pattern of characteristic thoughts, feelings, and behavior that persist across time and situations and distinguish one person from another.

HISTORY

Prior to the advent of scientific psychology, the study of individual differences in character was dominated by the ancient Greek theory of the four humors. The Greeks believed that four bodily fluids—blood, black bile, yellow bile, and phlegm—controlled the temperament. Hippocrates (460??–377 B.C.) argued that four different temperaments resulted from imbalances of the four fluids.

Prior to the twentieth century, there were various theories of character, as it was then called. Then, as now, the word, *character* suggested a moral dimension that the word *personality* did not. Nineteenth-century psychologists were not shy about investigating a moral dimension.

Beyond a reluctance to impose their own moral judgments on their subjects, early twentieth-century psychologists were also looking for grand theories of the commonalities amongst people. Studies of personality are studies of individual differences. It took a genius like Freud to develop a theory about common factors operating in different contexts to create individual differences.

CURRENT APPROACHES

The more exotic theories of personality have given way to typologies, a focus on what different sorts of personalities are actually out there, as opposed to explanations of why the different personalities are as they are. This approach is due to Gordon Allport, who found Freud's theories too elaborate. On the one hand, in the early stages, any science must discover what is out there before it can theorize as to why it is out there. On the other hand, descriptions of personality are bound to be influenced by the theoretical notions of the researcher. Simply classifying personality types means hiding those theoretical biases, rather than exploring them.

With the latest advances in genetics and brain scanning, the best hope for personality theory may be in finding biological correlates for the typologies that have been discovered.

Personality Theories

The theories of personality are hypothetical statements about the structure and functioning of individual personalities. They help us understand the structure, origins, and correlates of personality. They also help us predict behavior based on what we know about personality. The theories differ in their approaches because they begin at different points, have different sources of data, and try to explain different phenomena. Thus, each theory can teach us something about personality, and together they can teach us something about human nature.

FREUD'S PSYCHODYNAMIC THEORY

The psychodynamic theories of personality development are based on the work of Sigmund Freud, but have been modified by others over the decades. These theories are all concerned with powerful, but largely unconscious, motivations believed to exist in each of us. They also all maintain that personality is governed by conflict between opposing motives, anxiety over unacceptable motives, and defense mechanisms that prevent anxiety from becoming too great.

Freud argued that the events within a person's mind that motivate behavior are the core of personality. All behaviors are motivated, and all actions are determined by those motives. Our actions emerge from what we really desire, even when we do not know what it is that we want. Freud also argued that all mental and behavioral reactions, or *symptoms,* are determined by our early experiences. Rather than being arbitrary, these symptoms are meaningfully related to significant life events. Specifically, he believed that experience in infancy and early childhood had the most profound impact on personality formation and adult behavior patterns. He assumed a continuity of personality development from "the womb to the tomb."

Drives and Instincts

Each person is assumed to have inborn instincts or drives that are *tension systems* created by the organs of the body. When these energy sources are activated, they can be expressed in many different ways. For example, the sex drive can be expressed directly through sexual activity or indirectly through jokes or art. Freud argued that there are two primary drives. First, *self-preservation* is the drive to meet the needs of hunger and thirst. Second, *eros* is the driving force related to sexual urges and preservation of the species. It includes not only the drive for sexual intercourse, but also includes all other attempts to seek pleasure or to make physical contact with others. The urges stemming from these drives or instincts demand immediate satisfaction, whether through direct or indirect means.

Unconscious Processes

Freud held that the unconscious was the determinant of human thought, feelings, and action. He argued that behavior can be motivated by drives that we are not aware of consciously. We may act without knowing why or without direct access to the true cause of our actions. This is illustrated in Freud's notion of the structure of personality. He believed that personality differences arise from the different ways in which people deal with the urges that stem from their fundamental drives.

To explain these differences, Freud pictured a continuing battle between two antagonistic parts of the personality, the *id* and *superego*, and moderated by a third aspect of the self, the *ego*.

The id is made up of the primitive, unconscious part of the personality. It is the storehouse of the fundamental drives. It operates irrationally, acting on impulse and pushing for expression and immediate gratification. It does not consider whether or not what is desired is realistically possible, socially desirable, or morally acceptable. The id is driven by the *pleasure principle*, which is the unregulated search for gratification. It is governed by *primary-process thinking*, which is a form of thinking that is primitive, illogical, irrational, and fantasy-oriented.

The superego is the storehouse of an individual's values, including moral attitudes learned from society and parents, etc. It corresponds roughly to the common notion of conscience. The superego is driven by the *morality principle*, which insists on doing what is right.

The ego is the reality-based aspect of the self. It represents a person's view of physical and social reality or his conscious beliefs about the causes and consequences of behavior. Part of the ego's job is to choose actions that will gratify the id's impulses without undesirable consequences. The ego is driven by the *reality principle*, which puts reasonable choices before pleasurable demands. It is governed by *secondary-process thinking*, which is a form of thinking that is rational, realistic, and oriented toward problem-solving.

Thus, in Freud's view, personality is the constant conflict between the id and the superego, with the ego mediating the conflict. The id wants immediate gratification of its urges, but society, through the superego, may not allow that. This conflict is mediated by the ego, which seeks a compromise. Freud assumed that behavior is the outcome of this internal conflict.

For example, you are dining at a fancy restaurant and catch sight of the dessert tray. The id wants you to you dive at the tray and gobble down the desserts. The superego vetoes this strategy on the grounds that leaving the table and grabbing handfuls of chocolate cake is inappropriate. The ego finds a compromise and you call for the waitperson and ask to be shown the dessert tray. You make a selection and dine discreetly with a fork.

Ego Defense Mechanisms

Imagine constantly being caught between two powerful conflicting forces. You can imagine that you would be overwhelmed and possibly sustain some damage. You would find a way to protect yourself. Freud imagined that a person's ego was in just such a situation. He argued that we develop *defense mechanisms*. The defense mechanisms are mental strategies used to defend the ego in the daily conflict between id impulses and the superego's demands to deny them. Freud considered them vital to psychological coping with these powerful inner conflicts. Utilizing defense mechanisms allows a person to maintain a favorable self-image and to sustain an acceptable social image.

Freud further argued that we have two lines of defense against the conflict between the id and the superego. The first line of defense is *repression*. This occurs when extreme desires are pushed out of conscious awareness into the unconscious. If the id impulse is no longer in conscious awareness, then there is nothing for the superego to fight against and there is no conflict for the ego to mediate. It is the psychological process that functions to protect an individual from experiencing extreme anxiety or guilt about having ideas or impulses that are unacceptable or dangerous to express. However, repression does not always work.

The second-line defense mechanisms are triggered by anxiety, which signals that repression has failed and a conflict is about to emerge into consciousness. We can use any of the second-line defense mechanisms, or a combination of them, depending on the conflict we are trying to avoid. The most common defense mechanisms are listed here:

- **Displacement**   Discharging pent-up feelings, usually of hostility or frustration, on objects or people less dangerous than those that initially aroused that emotion. For example, instead of hitting your boss for yelling at you, you go home and yell at your dog. By displacing hostility, the id's impulse to yell back is satisfied without conflicting with the superego's knowledge that hitting the boss can get you fired.

- **Fantasy**   Gratifying frustrated desires in imaginary achievements or scenarios. For example, a child who was abandoned by a parent may fantasize that the parent is really a spy who had to leave for the child's security. By engaging in fantasy, the child's id impulse to be loved is no longer in conflict with the superego's knowledge that she was abandoned.

- **Identification**   Increasing feelings of self-worth by identifying with another person or institution with positive attributes. For example, a mediocre high-school football player may identify with a famous professional football player because they both wear the same jersey number. Through identification, the high-school player avoids the conflict between the id's impulse for high self-esteem and the superego's insistence that he be the best player by allowing him to "take on" the positive attributes of the professional player.

- **Isolation/intellectualization**   Separating incompatible thoughts or feelings into logic-tight compartments so that they are never thought of simultaneously or in relation to each other. For example, your significant other bears an uncanny resemblance to your parent, but you also have a strong sexual attraction to your significant other. Through isolation (or intellectualization), you never think of the resemblance when you are sexually attracted to your partner, nor do you ever think of being sexually attracted to your partner when you think of the resemblance.

- **Overcompensation**   Either covering up a perceived weakness by overemphasizing some other characteristic, or making up for a frustration in one area by over-gratifying oneself in another area. For example, a person may make up for a lack of sex by eating too much chocolate. By overcompensating with chocolate, the id's impulse for sexual gratification is not in conflict with the superego's rules on finding a sexual partner. Eating chocolate is socially acceptable.

- **Projection**   Placing blame for one's difficulties on others, or attributing one's own "forbidden" desires to others. Freud often discussed projection in therapeutic situations. For example, a patient may find himself attracted to his therapist. The id's impulse for gratification comes into conflict with the superego's knowledge that this is not an acceptable attraction. To resolve this conflict, the patient projects his attraction onto the therapist by thinking that the therapist is attracted to him.

- **Rationalization**   Attempting to prove that one's behavior is rational and justifiable and thus worthy of approval from the self and others. For example, a smoker may continue to smoke by rationalizing that something else will probably kill her first. The id's impulse for pleasure is no longer in conflict with the superego's insistence that smoking can cause disease.

- **Reaction formation**   Preventing dangerous desires from being expressed by endorsing opposing attitudes and behaviors and then using them as barriers. For example, a homosexual who is uncomfortable with his or her homosexuality may resolve this conflict by gay-bashing. The conflict between the id's impulse for sexual gratification and the superego's stigma of homosexuality is resolved by engaging in thoughts and behaviors consistent with the superego.

- **Regression**   Retreating to earlier developmental levels involving more childish responses. For example, when a new infant is brought into the family, the older child may start wetting the bed and sucking his thumb. The id's impulse for attention is in conflict with the superego's requirement that he share his parents' attention. By acting more infantile, and likely getting the attention he seeks, the conflict is resolved.

- **Sublimation**   Gratifying frustrated sexual desires in substitutive nonsexual activities that are socially acceptable. For example, a person can work off sexual frustration by training for a marathon. The logic is similar to overcompensation, but sublimation is restricted to sexual frustration.

- **Undoing**   Trying to atone for an unacceptable desire or action. For example, an unscrupulous businessperson may contribute a large sum of money to charity to "make up" for his shady business practices. By doing a "good deed," he resolves the conflict between the id's impulse to make money and the superego's morality.

- **Denial of reality**   The defense mechanism of last resort. Protecting the self from unpleasant reality by refusing to perceive that reality. For example, an addict may refuse to believe that she has an addiction. By denying the reality of the addiction, the id's impulse to use the substance does not conflict with the superego's insistence that addiction can be dangerous and socially unacceptable.

Critical Caution

Denial is commonly referred to in "pop" psychology through television and film, particularly in relation to addiction. Real denial is much less common. It is important to note that the popular definition is much broader than what Freud intended and much broader than what professional psychologists intend. Psychologists do not use it as a metaphor for lying, nor do they use it as a substitute for stubbornness or resistance to treatment. You have to be careful when accusing someone of being in denial, as there is no defense to that accusation. ("You're in denial!" "No, I'm not!" "See! I told you so!") Such banter may be amusing, but it has nothing to do with the defense mechanism of denial.

Bio Bites: Real Denial

One of the authors of this book thought he knew what denial was—until he had a chance to see the real thing. A male acquaintance in his 70s, with a long history of good health, had his first heart attack far from any hospital. He was seen in an emergency room several hours later, but was only correctly diagnosed days later and did not get to a proper hospital until over a week afterwards. Once he arrived at the hospital, the diagnosis was confirmed and he was immediately scheduled for quintuple bypass surgery.

After the successful surgery, recovering in the cardiac care unit, the man was concerned. "I'm still not sure I had a heart attack," he said.

"Why not?"

"The pain I had that first day was nowhere near my heart."

"Where was the pain?"

"Across my back, like a band."

"Where was this band?"

"At the height of my underarms."

The author then lifted his own arms to see if there was anywhere on his own back at the height of his own underarms that was more than a few inches from his own heart. He looked back at the man lying in the hospital bed and decided to let the matter rest. This man was convinced he had not had a heart attack, even if he had to relocate his heart to do it. That is denial.

Psychosexual Personality Development

Freud argued that at each stage in a child's life the drive for pleasure centers around a particular part of the body. He believed that adult personality is shaped by the way in which the conflicts between these early sexual urges (id-driven) and the requirements of society (superego-driven) are resolved. The failure to resolve any conflicts can result in a *fixation* in which a person becomes stuck in that conflict.

Big Background: Childhood Sexuality

Freud received a great deal of criticism for his psychosexual stages of personality development. He was the first to posit that sexuality was not merely something that appeared in adulthood. Children, and even infants, strive for pleasure. While Freud's theory of personality development did not lend itself to scientific investigation, the idea that sexuality begins before adulthood has been supported.

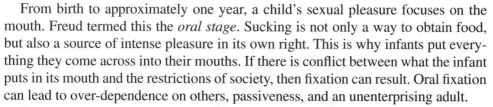

From birth to approximately one year, a child's sexual pleasure focuses on the mouth. Freud termed this the *oral stage*. Sucking is not only a way to obtain food, but also a source of intense pleasure in its own right. This is why infants put everything they come across into their mouths. If there is conflict between what the infant puts in its mouth and the restrictions of society, then fixation can result. Oral fixation can lead to over-dependence on others, passiveness, and an unenterprising adult.

From one to two years of age, children enter the *anal stage*. This stage coincides with a child's physiological ability to gain voluntary control over his bowel movements. As a result, Freud argued that the child derives pleasure from controlling these actions. Nevertheless, the demands of toilet training can impose a conflict between id impulses for pleasure and society's rules for when and where this is appropriate. Fixation can result in extreme orderliness and excessive neatness. Indeed, this is where the characterization of someone being "anal" is derived.

From three to six years of age, children are in the *phallic stage*. During this stage, pleasure is centered on the genitalia and, specifically, masturbation. Freud argued that the Oedipus/Electra conflict (Oedipus for males, Electra for females) is the task to be resolved at this stage. This conflict involves an intense desire to take the place of the same-sex parent in the affections of the opposite-sex parent. The child is also afraid that the jealous same-sex parent will seek revenge. Hence, the conflict. If not resolved, male children experience castration anxiety, and female children experience penis envy. The healthy resolution is one in which the child recognizes that she can never possess the opposite-sex parent, so instead the child tries to become like the same-sex parent. This identification process is critical to the development of the child's superego.

From six to thirteen years of age, children enter the *latency period*. Freud argued that children are busy learning cognitive and social skills. Any sexual impulses appear to remain in the background. Finally, during adolescence, children enter the *genital phase*. The focus is on the pleasures of sexual intercourse. Conflict arises between their sexual desires and society's constraints on appropriate sexual behavior.

Final Notes on Freud

Freud's work has received a great deal of criticism, some deserved and some undeserved. While many students tend to giggle through most descriptions of Freud's work, it is important to not dismiss it entirely. It is true that much of Freud's work on personality is descriptive. This means that his theory is not testable using the scientific method. However, this does not mean that Freud's descriptions are not accurate. For example, his descriptions of the defense mechanisms are easily demonstrated in a scientific setting. What is not clear is whether or not these mechanisms are protecting an "ego" or working via some other process. Further, his descriptions of the psychosexual stages led to empirical work in human sexual

Reading Rules

These issues about Freud's theories (or Freud's ideas when they are not testable) make an important point about science. The value of a scientific theory is rarely in its being right. The overwhelming majority of scientific theories are eventually shown to be wrong, at least in some sense. The importance of a theory is in how it advances science by forcing its opponents to test it in order to prove it wrong. Very few of Freud's ideas survived scrutiny, but that scrutiny had important influence on almost every aspect of scientific psychology.

For example, proving the existence of a place in the mind called the unconscious may be impossible. On the other hand, the fact that Freud suggested that we ourselves are not always aware of what is going on in our own minds made psychologists aware that they needed a better way of investigating the mind. Modern psychological experiments are the result.

development that correctly included childhood. Again, his views of the process may not be correct, or even testable, but it did pave the way for later scientific work that incorporated his descriptions.

POST-FREUDIAN PSYCHODYNAMIC THEORIES

Much of the post-Freudian psychodynamic theories of personality focused on various aspects of Freud's overall theory. Carl Jung sought to expand the notion of the unconscious, and Alfred Adler argued against Freud's focus on sexual conflicts.

Jung's Analytic Psychology

Jung focused his description of personality on the importance of the unconscious. He argued that there are two layers of the unconscious, the *personal unconscious* and the *collective unconscious*. The personal unconscious is equivalent to Freud's notion of the unconscious. It contains material that has been repressed or forgotten and therefore is not in conscious awareness. The collective unconscious is the storehouse of latent memories inherited from people's ancestral past. Jung argued that the entire human race shares the same collective unconscious. It consists of archetypes that we all recognize and understand on an unconscious level. These archetypes are ancestral memories and not memories of events or experiences. They are emotionally charged images and forms of thought that have universal meaning. They manifest themselves in our dreams and in a culture's art, literature, and religions.

Adler's Individual Psychology

Adler devised his notion of personality as a response to Freud's focus on sexual conflicts. Adler argued that people have a universal drive to strive for superiority. We try to improve ourselves, adapt, and master life's challenges. This, and not sexual pleasure, is believed to be the primary motivation for behavior. He further argued that failing to meet our goals can cause normal feelings of inferiority. We try to compensate for these feelings by developing our abilities. However, excessive feelings of inferiority caused by parental pampering or parental neglect can cause an inferiority complex that interferes with the normal processes of striving for superiority. Instead of working to master life's challenges, people with an inferiority complex try to overcompensate in other areas. They may try to gain power, achieve status, and acquire the trappings of success.

BEHAVIORIST THEORIES

The behaviorists view personality in terms of learned behaviors. Instead of creating specialized personality theories, they expanded the principles of learning theories to explain the differences in personality. To review the behaviorist's theories in detail, refer to Chapter 11.

Operant Conditioning

B. F. Skinner and other operant conditioning theorists argued that personality development is based on behaviors that are either rewarded or punished in the environment. Behavior varies only as the contingencies in the environment vary. Therefore, what others call "personality" may actually just be behavioral patterns sustained by a long-term stable environment.

For example, the way we respond to a social situation is a function of our past experiences with similar social situations. We can either start talking to new people, stay near the people we already know, or try to find a way to get out of the social situation. What we tend to do is based on whether or not we were rewarded (reinforced) for a particular behavior in the past. If we were rewarded for talking to new people in the past, then we are more likely to talk to new people in the present situation. However, if we embarrassed ourselves, then we are less likely to repeat the behavior in the present social situation. Therefore, what personality theorists refer to as personality are really behavioral tendencies based on a long history of contingencies in similar circumstances.

Even when circumstances change, our tendency to avoid unpleasant situations may prevent our behavior from changing. We may have matured past the point of embarrassing ourselves in public. Our peers may have outgrown their tendency to make fun of us. However, our punishing experience with past opportunities to meet new people may prevent us from taking advantage of these new and better opportunities. Such people may be termed "shy." Operant theorists would say they are under control of contingencies that are no longer in effect.

Social Learning Theory

Albert Bandura and the social learning theorists argued that personality develops from imitating the behavioral tendencies we observe in others. Imitation is particularly influential if we observe a behavior leading to a positive outcome. For example, if we observe someone being assertive in a job interview, we are likely to be assertive ourselves. If we also observe that the other person got the job and was therefore rewarded for the assertive behavior, then we are even more likely to be assertive. The social learning theorists argued that a person's *self-efficacy*, or our belief in our ability to perform behaviors that lead to expected outcomes, influences our behavior. If our self-efficacy is high, we feel confident that we do the necessary things to earn reinforcers. When our self-efficacy is low, we worry that the necessary responses are beyond our abilities. This self-efficacy influences what challenges we take on and how well we perform these tasks. Thus, what others view as personality is believed to be response patterns that are learned by receiving contingencies from similar situations, by watching others receive contingencies, and by our self-efficacy.

Since the cognitive revolution, cognitivists have adopted Bandura's ideas as their own. The social-cognitive perspective has continued, expanding on the early ideas of Bandura and others. An important discovery is the role of *locus of control* in personality. People who view their own behavior as being more effective in delivering desirable outcomes are said to have an *internal locus of control*. Those who view their own behavior as less effective are said to have *external locus of control*. Internal versus external locus of control correlates with such personality traits as perseverance and optimism (both justified and unjustified).

Trait Theories of Personality

Trait theories are very different sorts of theories than the other theories of personality. The focus of trait theories is not so much on why people are as they are, but what sorts of people are out there. Trait theories are more classification systems than true theories. Classification systems play a vital role in science. We need to know what is out there before we can understand why it is there or how it got to be there.

Each of the trait theories of personality proposes hypothetical continuous dimensions that vary in quality and degree. Traits are generalized action tendencies that people possess in varying degrees and lend coherence to a person's behavior in different situations and over time.

There are two general ways to think about traits, both of which are determined by measurement techniques. The first way is to think of a trait as a dichotomous continuum in which opposites sit at either end. For example, a bipolar "introversion-extraversion" trait scale would put introversion on one end and its opposite, extraversion, on the other. Which side of the scale a person falls upon determines whether he or she is introverted or extraverted and to what degree.

Another way to both measure and think about a personality trait is to imagine the trait as a single continuous scale with "extremely" and "not at all" anchoring the scale at either end. This allows you to think about, and measure, introversion and extraversion separately. While these traits may be highly negatively correlated, they can also be considered different traits. Everyone has an introversion score, which may be high, and an extraversion score, which may be low. Thus, we can think about oppositional traits individually.

ALLPORT

Gordon W. Allport viewed traits as the building blocks of personality and the source of individuality. He held that traits produce coherence in behavior because they are enduring attributes and they are general in scope. They connect and unify a person's reactions to a variety of situations. Thus, they relate sets of stimuli and responses that, at first, seem to have little to do with each other.

What is original about Allport's theory is the way he organized personality. He proposed a hierarchy of traits that form the structure of personality. At the top of the hierarchy are the *cardinal traits*, which are those traits a person organizes his or her life around. Below that are the *central traits*, which represent the major characteristics of a person. Finally, at the bottom of the hierarchy, are the *secondary traits*. These are specific personal features that help us to predict an individual's behavior, but are less useful for understanding that person's behavior. For example, Mother Theresa's cardinal trait might have been self-sacrifice for the good of others. From this, we can predict several central traits, such as honesty, altruism, or optimism. Her secondary traits might have involved personal things we knew about her, such as food preferences. These preferences might have allowed us to predict what she would have ordered for lunch, but they do not give us much insight into her personality.

Allport argued that this personality structure, rather than environmental conditions, is the critical determinant of a person's behavior. The same situation can have different effects on different people. Although he recognized common traits that people in a given culture share, he was most interested in discovering the traits

that make each person unique. He believed that common and unique traits were combined in different ways, resulting in unique individuals.

EYSENCK

Hans Eysenck also proposed a model of personality that is structured hierarchically. His model links personality types, traits, and behaviors into a hierarchical system. However, he argued that personality is largely genetic. Eysenck's hierarchy begins at the bottom with *single responses*, or single actions or cognitions. The second level consists of *habitual responses*, or several actions or cognitions that tend to occur regularly. Several correlated habitual responses form *traits* and the third level of the hierarchy. Finally, correlated traits form *types*, and the top level of the hierarchy.

THE BIG FIVE

One hazard of the trait theories of personality is that different researchers tend to study different traits. They also design many different ways to measure those traits and sometimes create their own names for the dimensions they measure. This creates confusion when psychologists want to make generalizations across many collections of data. This problem led to a search for common dimensions of meaning that would link the wealth of information together. The goal of this search was to develop a common language that all personality psychologists could use to compare their results. Ideally, such a system would be comprehensive, easy to understand,

---

Tips on Terms: Traits and Types

The practice of classifying personalities into types goes back to the ancient Greeks. The different humors represented different traits. Predominance of one trait corresponded to one character type. More modern type theorists use statistical methods to identify patterns of traits that correspond to each personality type, but the basic concept is the same. *Traits* are collections of related behaviors. *Types* are collections of people who exhibit common traits. Type theories remain very popular. Some, such as Eysenck's, are attempts to classify all people into a universal system of types. Some, such as the Type A theory of personality, are attempts to find a specific personality type that is a useful predictor of something important—in this case, risk of heart disease.

Table 5-1   The "Big Five" Personality Traits

Trait	Score High/Positive Correlation	Score Low/Negative Correlation
Openness	Creative, intellectual, open-minded	Simple, shallow, unintelligent
Conscientiousness	Organized, responsible, cautious	Careless, frivolous, irresponsible
Extraversion	Talkative, energetic, assertive	Quiet, reserved, shy
Agreeableness	Sympathetic, kind, affectionate	Cold, quarrelsome, cruel
Neuroticism	Stable, calm, contented	Anxious, unstable, temperamental

independent of any particular theory, and flexible enough to allow for different levels of analysis.

This research resulted in what are known as the "big five" personality traits, or the *Five Factor model*. Researchers found that each variation of personality scales could be categorized into five factors. Most of these traits should be familiar, as they are commonly used to describe personality outside of psychology. They are listed in Table 5-1, as well as the types of personality scales that a person high and low on the trait would score.

BIOLOGICAL CONNECTIONS

Eysenck's speculation that genetics may play an important role in personality represents a long tradition of concern about the resilience of personality in the face of learning and experience. Even young infants seem to have differences in character or at least in temperament. Do people come genetically prewired to trend in one personality direction or another?

Empirical evidence for a genetic component to personality has not been strong. However, with the completion of the mapping of the human genome, new ways of studying this question are now available. One fascinating line of research on something called the *serotonin transporter polymorphism* can give us an idea of what this very new sort of research may look like.

Serotonin is a very important brain chemical. Another brain chemical, called the serotonin transporter (5-HTT), regulates the amount of serotonin in the brain. There are two common types of serotonin transporter, the long and the short, determined by a polymorphism in the controlling gene. Early studies indicate that some personality characteristics, such as neuroticism, as measured by trait-based scales, correlate with the genetic profiles of people's 5-HTT gene site. While the effects are small, the fact that we can test people's DNA profile for specific genetic differences in brain chemicals and compare those DNA results to measures of behavior opens up a fascinating new world of personality research.

Summary

Personality research focuses on the what and the why of individual differences. The older tradition of psychodynamic and behavioral theories focused on why different people behave differently. The notion of traits, enduring trends in behavior that persist across situations and are measured in degrees, is central to the newer approach of categorizing what personality types can be found. These traits can then be correlated to other behaviors of interest. New methods in DNA research may make it possible to find a biological basis for personality traits.

Quiz

1. Personality is a pattern of characteristic _____ that persist across time and situations and distinguish one person from another.

    (a) Thoughts

    (b) Behaviors

    (c) Feelings

    (d) All of the above

2. Which of the following are important influences in the psychodynamic theories of personality?

    (a) Early experiences

    (b) Unconscious motivations

    (c) Both a and b

    (d) None of the above

3. The _____ is made up of the primitive, unconscious part of the personality.

    (a) Id

    (b) Ego

    (c) Superego

    (d) Drive

4. The _____ is driven by the morality principle.

   (a) Id

   (b) Ego

   (c) Superego

   (d) Personality

5. _____ is the discharge of pent-up feelings, usually of hostility or frustration, on objects or people less dangerous than those who aroused that emotion initially.

   (a) Denial

   (b) Displacement

   (c) Isolation

   (d) Projection

6. According to Jung, the _____ unconscious consists of archetypes that we all recognize and understand on an unconscious level.

   (a) Collective

   (b) Personal

   (c) Rational

   (d) All of the above

7. _____ theorists argued that personality development was based on response tendencies that are either rewarded or punished in the environment.

   (a) Trait

   (b) Psychodynamic

   (c) Social learning

   (d) Operant conditioning

8. The _____ theories of personality propose hypothetical continuous dimensions that vary in quality and degree.

   (a) Psychodynamic

   (b) Classical conditioning

   (c) Trait

   (d) Social learning

9. According to Allport, _____ traits are those traits a person organizes his or her life around.

   (a) Secondary

   (b) Cardinal

   (c) Primary

   (d) Central

10. Eysenck's idea that genetics determines personality may be supported by studies correlating behavior with _____.

   (a) Brain chemicals

   (b) DNA profiles

   (c) Brain scans

   (d) None of the above

CHAPTER 6

Health Psychology

Our psychology has a surprising amount of influence on our health. Psychologists study these influences to help improve our health.

Overview

Psychology and health are interrelated in two basic ways. First, there are the bad behaviors we indulge in that damage or endanger our health, as opposed to the good behaviors we all imagine we could be doing that would benefit, or at least not risk, our health. Second is *stress*. Stress is a psychological phenomenon that can have an enormous negative impact on our health. Our response to stress is *coping*. Good coping behaviors can enable us to deal with stressful situations with less stress.

HISTORY

Long before psychology and medicine intersected, people noticed that their health affected their behavior and their behavior affected their health. Centuries before medicine became a profession and millennia before psychology developed as a science, the idea that what we do can affect how we feel has influenced the progress of both fields.

Dating back to approximately 400 B.C., health "theorists" recognized that good health and ill health are different states of being. They further recognized that there are things we can do to change our state of being. What early theorists recommended depended on what they believed the cause of illness to be. For example, the earliest theorists believed that illness was caused by the presence of evil spirits in the body. In order to cure the illness, they often drilled a hole in the skull to allow the spirits to escape. By the fourth century B.C., Greek theorists believed that the balance of humors (fluids) in the body affected health. If the humors were out of balance, illness resulted. To restore health, specific changes to the diet were prescribed.

Psychology and medicine did not come together until Freud's work on unconscious conflicts. As a trained physician, he noticed that some illnesses could not be traced to a biological cause. Instead, he argued that these unconscious conflicts caused physiological symptoms. Thus, psychological factors could affect physical health. To restore health, the physician must treat the underlying psychological conflict.

CURRENT APPROACHES

As psychology and medicine progressed, the theories behind the relationship between mental processes and health became more sophisticated. The rise of the *biopsychosocial* model of health is an example of this point. This model recognizes that health (or lack thereof) is determined by a combination of biological, psychological, and environmental factors. In order to treat illness, or to maintain good health, each type of factor must be considered. Acceptance of the biopsychosocial model spurred the rise of health psychology as a formal discipline.

Another contributing factor to the rise in health psychology was the changing patterns of illness and mortality in the twentieth century. It used to be that infectious diseases, such as pneumonia, were the leading killers. Today, the leading causes of death are heart disease and cancer. Once we developed the medical capability to cure infectious diseases, preventable illness became a leading cause of mortality. Thus, changes in our behavior can prevent what are now the leading causes of death. It should come as no surprise that the surge in health psychology research coincided with this new trend.

Theories of Health

Psychologists interested in health have developed theories that attempt to identify the causes of good and bad behavior related to health. Knowing more about what causes health-related behavior helps us develop systems for promoting more healthy behaviors.

HEALTH BELIEF MODEL

The biopsychosocial model of health has given rise to several theories of health behavior. The oldest is the Health Belief Model, which states that four factors determine whether or not individuals will alter their behavior. These factors are outlined here:

- **Perceived susceptibility**   How likely a person believes they are to contract a disease or experience a health threat. The more susceptible a person believes himself to be to a health threat the more likely he is to try to alter his behavior.

- **Perceived severity**   How disruptive a person believes the health threat to be. The more disruptive the illness is believed to be, the more likely a person is to alter her behavior.

- **Benefits and barriers**   A cost-benefit analysis to determine whether the barriers to changing behavior are outweighed by the benefits. If the benefits outweigh the barriers, a person is more likely to alter his behavior.

- **Cues to action**   Extraneous factors, such as social support, socioeconomics, age, or advertising (for example, public service messages). The more cues to action a person is exposed to, the more likely she is to alter her behavior.

According to this model, people are likely to change their behavior if their perceived susceptibility is high, the perceived severity is high, the benefits outweigh the barriers, and there are positive cues to action. For example, the risk of contracting HIV decreases greatly when people use condoms when engaging in sexual behavior. However, it has proven rather challenging to encourage people to change their behavior to use condoms on a regular basis. The target health behavior is the use of condoms. What, according to this model, would have to happen for an individual to change this target behavior?

First, a person would have to believe that he was at a high risk of contracting HIV through sexual activity. Second, the person would have to believe that HIV's impact on his quality of life would be disruptive. Third, he would have to believe that the cost or inconvenience of purchasing and using condoms is less than the benefits of the protection. Finally, his cues to action may involve encouragement from his sexual partner, peer pressure, or exposure to frequent ads promoting condom use, among others.

PROTECTION MOTIVATION THEORY

The Protection Motivation Theory extends the Health Belief Model. One problem with the earlier theory is that, although good at predicting attempts to *alter* behavior, it does not always predict success. The Protection Motivation Theory adds the "self-efficacy" factor. Self-efficacy is people's belief in their ability to successfully alter their behavior. When their self-efficacy is high, they are more likely to attempt to alter their behavior and are more likely to succeed.

THEORY OF REASONED ACTION

The Theory of Reasoned Action focuses on the influence of behavioral intentions on behavioral change. This theory argues that the intention to alter behavior is influenced by the *attitudes* and subjective norms surrounding that behavior. Attitude refers to a belief that the behavioral change will bring about some outcome and that the outcome is desirable. For example, if condom use prevents HIV exposure and that is perceived to be a favorable outcome, then the intention to use condoms increases. Subjective norms refer to the perception of how other people will judge the behavior and how much a person wants to comply with others' wishes. To continue the example, if a person's partner views condom use favorably and the person is highly motivated to comply with his partner's wishes, then his behavioral intention will be high. With positive attitudes and strong subjective norms, behavioral intention increases and the person is more likely to change his behavior.

THEORY OF PLANNED BEHAVIOR

Much like the earlier theories, the Theory of Reasoned Action did not take self-efficacy into account. The Theory of Planned Behavior is an extension of the Theory of Reasoned Action that includes this factor. Thus, if a person believes that he can successfully alter his behavior, then he is more likely both to try and to succeed.

Health-impairing Behavior

Much of the time, the behaviors we engage in on a daily basis do not threaten our life. Eating that ham sandwich for lunch or having a glass of wine with dinner typically will not kill us. Or will it? For some of us, these small, seemingly innocuous behaviors can add up to a pattern of behaviors that do have health implications.

Advances in medicine have provided us with many ways to achieve and maintain good health. Why is it, then, that so many of us jeopardize our health, and possibly our lives, by engaging in unhealthy behaviors? Why do we smoke, abuse alcohol, neglect proper nutrition, avoid exercise, or engage is risky sexual behaviors? The illnesses caused by smoking, obesity, alcohol abuse, and sexually transmitted diseases are largely preventable. Yet, some of us continue to put ourselves at risk.

The theories of health behavior outlined previously help us answer this question. Earlier, we discussed each theory in terms of what is required to increase the likelihood of changing risky behavior. In this section, we outline each theory in terms of what leads us to continue engaging in risky behavior:

- Health Belief Model
  - Low perceived susceptibility to the health threat
  - Low perceived severity of the health threat
  - Barriers to behavior change outweigh the benefits
  - Few cues to action
- Protection Motivation Theory
  - Low perceived susceptibility to the health threat
  - Low perceived severity of the health threat
  - Barriers to behavior change outweigh the benefits
  - Few cues to action
  - Low self-efficacy of behavior change
- Theory of Reasoned Action
  - Negative or neutral attitude toward the outcome of behavior change
  - Negative or neutral subjective norms
- Theory of Planned Behavior
  - Negative or neutral attitude toward the outcome of behavior change
  - Negative or neutral subjective norms
  - Low self-efficacy of behavior change

PATIENT BEHAVIOR

Why do we engage in risky or unhealthy behaviors? In most cases, it is because the behavior brings us some form of pleasure. Fatty foods taste good. Smoking can be relaxing. In the short term, engaging in these behaviors makes us feel better. In turn, not engaging in healthy behaviors can also bring pleasure by making our life easier. Not exercising requires less effort than exercising. Cooking takes more time (although usually less money) than hitting the fast food place. Therefore, when we discuss changing our behavior, we are actually talking about denying ourselves some form of pleasure.

We have examined what factors influence the decision to maintain or change behaviors. What situations influence those factors? In this section, we will examine the individual and familial-cultural barriers that prevent us from maintaining good health or preventing illness.

Individual Barriers

There are two individual conditions that influence the decision to engage in risky behaviors. The first is the amount of knowledge a person has about the risks of her behavior. In the 1930s and 1940s, people did not know what damage smoking could cause. Today, that is hardly the case. Smoking prevention programs are routinely taught in schools, public service announcements are shown daily on television, and tobacco companies have been sued successfully by some states to recoup the healthcare costs spent on smoking-related illnesses. The cigarette packages themselves even have warnings on them.

We also have ample information pertaining to the importance of healthy eating and exercise, the use of condoms to prevent contracting sexually transmitted diseases, and the effects of alcohol overuse. One would be hard-pressed to argue today that lack of knowledge contributes to our unhealthy behavior. However, willing ignorance remains a problem. It is not that we do not know about the threats to our health, but that we choose to ignore them.

A second, more insidious problem is that many unhealthy behaviors begin in adolescence. The sense of invulnerability that teens experience makes them particularly susceptible to developing unhealthy behavioral habits. They may have the knowledge of the consequences of risky behavior, but they do not think of their behavior in terms of what may happen thirty years in the future. The time lag between the immediate rewards brought by unhealthy behavior and the possible negative consequences far in the future is so great as to negate the effect the negative consequences may have on decision making.

Family Barriers

Although unhealthy behavior patterns typically begin in adolescence, they are learned much earlier. Children of smokers tend to smoke. Children of obese parents tend to be obese. Children of alcohol abusers tend to abuse alcohol as well. There are two familial components that influence our tendency to engage in unhealthy behaviors. The first component is genetics. The tendencies to smoke, to be alcoholic, or to be obese do have genetic components. However, genetics cannot account for all of it. The second component involves learning. Young children tend to model, or mimic, the behavior they see around them. Thus, most of the behavioral patterns we learn are learned from our parents (or primary caregivers).

If our parents smoke, are alcoholic, or are obese, are we doomed? No. Adopting unhealthy behaviors via one's parents is not a foregone conclusion. Both genetics and learning increase our *tendency* to engage in these unhealthy behaviors and the *likelihood* that we'll experience the negative health consequences. Children of obese parents are more likely to have the genetic tendency for heart disease, high blood pressure, or a myriad of other health problems associated with obesity. However, if they never learn the unhealthy eating and exercise patterns, they are less likely to become obese in the first place. Thus, learning and genetics do not doom us to a lifetime of unhealthy behaviors. They just increase our tendency to engage in these behaviors, which makes unlearning them later in life more difficult.

Stress

Unhealthy patterns of behavior are not the only factors that influence our overall health. The stress we experience in everyday life also has a profound effect on our physical and psychological health. In this section, we will examine the nature of stress, how we respond to stress, and how stress can affect our health.

NATURE OF STRESS

It is easy to think of stress as a reaction to a major traumatic event, such as a natural disaster or combat experience. For most of us who are fortunate enough to not experience such major events, we still experience stress. Health psychologists define *stress* as the reaction to stimuli that threaten our physical or psychological well-being. The stimulus, or *stressor*, can be any object, event, or situation that is perceived as a threat to our equilibrium. What constitutes a stressor varies widely from person to person, but the stress response is relatively consistent.

The types of stress can be categorized by how we experience the stimuli emotionally. Thus, encountering the stimuli causes a change to our well-being, and we experience some emotional state. The four types of stress are listed here:

- **Conflict**   Conflict is caused by two or more incompatible behavioral options competing for expression. For example, choosing between a dead-end job and the prospect of prolonged unemployment can cause conflict.

- **Frustration**   Frustration occurs when a goal is thwarted. For example, getting caught in traffic when you are in a hurry can cause frustration.

- **Change**   Although not quite an emotion, change refers to those life events that require significant readjustment. For example, graduating from college, getting married, moving to a new city, and so on can all demand important changes to one's behavior patterns.

- **Pressure**   Pressure involves demands or expectations to behave or perform in a specific manner. For example, by expecting to earn an A in your psychology class, you are putting pressure on yourself to perform to a certain standard.

RESPONSES TO STRESS

We can think of our reaction to stress as having three components. We experience a physiological reaction, an emotional reaction, and a behavioral reaction.

Physiological

Our physiological reaction to stress is typically the same regardless of the stressor. There may be slight variations in the reaction's intensity, but the response follows the same physiological pattern. According to the theory of the *general adaptation syndrome*, we progress through three stages of physiological response to stress.

When we first recognize the stressor as a threat, our body experiences an alarm reaction. This is the flight-or-fight response in which the body prepares to cope with the threat. We experience an increased heart rate and higher blood pressure. Our breathing quickens, our perspiration increases, and our pupils dilate. Each of these reactions prepares the body to fight the threat or to flee from it.

Then comes the second stage. If the threat of the stressor continues (akin to chronic stress), our body begins to adapt to the threat. The physiological changes, while still higher than normal, begin to stabilize. Finally, if we cannot overcome the stress, we enter the third stage, exhaustion. Our body's resources are being depleted, our physiological reactions are decreasing, and our resistance to disease and illness increases. Our bodies cannot fight much of anything, including new threats.

Emotional

The relationship between a stressor and the emotional response it triggers is not as straightforward as the relationship between a stressor and the physiological response. However, based on the perceived level of threat brought by a stressor, some emotions are more likely to be triggered than others. For example, frustration is likely to bring about anger, whether it be a mild annoyance or an uncontrollable rage. Stress from life changes can lead to anxiety and fear.

Behavioral

The behavioral response to stress is termed *coping*. It is the deliberate effort to manage stress by overcoming it, reducing it, or tolerating it. Coping can be either adaptive or maladaptive. We typically think of the term *coping* as something inherently positive. If someone is coping with a problem, we generally think of this in terms of healthy behaviors. However, coping can also refer to maladaptive behaviors. The person coping with a problem may be coping poorly.

Coping mechanisms are the behavioral patterns, or coping strategies, we use to manage stress. Although people cope with stress in a variety of ways, we each develop certain styles, or patterns of behavior. Psychologists have identified three general types of coping mechanisms.

Those strategies that are *problem focused* attempt to cope with the stress by changing the stressful situation. This may involve trying to avoid stressful situations by planning ahead, asking for help in changing the stressful situation, or removing the stressor. Problem-focused coping works when we have some control over the situation.

Cognitive-focused strategies focus on changing the thoughts about the stressful situation. If a stressful situation is unavoidable, then we may try to change the way we perceive the stressor. This can involve both adaptive and maladaptive mechanisms. We may try to reevaluate the situation in order to put it in a more realistic light. On the other hand, we may use drugs or alcohol in order to block out thoughts about the stressor.

Finally, *emotion-focused strategies* attempt to alter the unpleasant emotions caused by the stressful situation. A strategy such as deep breathing to reduce anxiety is an example of this kind of coping. Again, alcohol or drugs may be used to escape the emotional distress. However, adaptive strategies are also used, such as engaging in activities that bring you pleasure or seeking sympathy from others. Emotion-focused coping works when the situation is out of our control, leaving us the option of managing our responses to it.

Fun Facts

A research study from the 1960s illustrates the profound effect stress can have on health. Before embarking on a two-year mission on an aircraft carrier, all crew members were surveyed by psychologists. After their return, their medical records were examined. It was found that sailors who had experienced one or more life-stressor events, such as marriage, divorce, birth of a child, death of a relative or close friend within 18 months of sailing had a higher rate of almost every kind of medical problem, from cancer and heart disease to ulcers to broken legs from falls.

STRESS AND HEALTH

Stress-related illness is becoming all too common in modern Western societies. Longer hours at work, fewer hours to relax, being constantly attached to our cell phones, and multitasking are just a few of the many reasons we experience more daily stress than our ancestors did just a few generations ago. The increased opportunity for stress has taken a toll on our collective health. Here, we have outlined some of the more common stress-related illnesses:

- **Heart disease**   Primarily caused by the increased blood pressure brought about by stress.

- **Cancer**   Although there is no known causal link between stress and cancer, stress can cause increased tumor growth.

- **Immune functioning**   The decreased immune functioning caused by stress can leave us vulnerable to bacterial and viral diseases, as well as decrease our ability to fight off any illness.

Summary

Health psychology is an important example of *applied psychology*. Most of the psychology discussed in *Psychology Demystified* is basic science. But there are many, many areas where psychology is applied to make our lives better. Health psychology is a preeminent example.

Quiz

1. The biopsychosocial model of health recognizes that health (or lack thereof) is influenced by _____ factors.

    (a) Biological

    (b) Psychological

    (c) Environmental

    (d) All of the above

2. The Protection Motivation Theory adds which factor to the Health Belief Theory of health behavior?

    (a) Self-efficacy

    (b) Genetics

    (c) Cues to action

    (d) Perceived severity

3. The theory of _____ focuses on the influence of behavioral intentions on behavioral change.

    (a) Biopsychosocial

    (b) Protection Motivation

    (c) Reasoned Action

    (d) Health Belief

4. Which of the following is an individual barrier to engaging in healthy behaviors?

    (a) Ignoring information about negative consequences

    (b) Not having enough information about negative consequences

    (c) Both a and b

    (d) None of the above

5. _____ is a type of stress caused by two or more incompatible behavioral options competing for expression.

    (a) Conflict

    (b) Frustration

    (c) Change

    (d) Pressure

6. According to the general adaptation syndrome, which of the following is not a step in the stress response?

   (a) Recognition

   (b) Adaptation

   (c) Exhaustion

   (d) Flight

7. _____ coping strategies attempt to cope with the stress by changing the stressful situation.

   (a) Emotion-focused

   (b) Problem-focused

   (c) Cognitive-focused

   (d) Biologically focused

8. Stress can lead to which of the following ailments?

   (a) Cancer

   (b) Reduced immune functioning

   (c) Heart disease

   (d) All of the above

9. According to the Theory of Planned Behavior, which of the following leads us to continue to engage in unhealthy behaviors?

   (a) Negative or neutral attitude toward the outcome of behavior change

   (b) Negative or neutral subjective norms

   (c) Low self-efficacy of behavior change

   (d) All of the above

10. Unhealthy behaviors are typically learned during _____.

   (a) Infancy

   (b) Childhood

   (c) Adolescence

   (d) Adulthood

CHAPTER 7

Psychological Disorders

Many people have some difficulty coping in the world. Some people have great difficulty coping in the world. Psychology has always taken a strong interest in understanding how and why people cope well or poorly. Following the *medical model,* many psychologists have classified the difficulties people have into various *psychological disorders*, akin to medical disorders. In this chapter, we discuss the various types of psychological disorders, the classification system, and the reasons for focusing attention on some disorders more than others.

Overview

The psychologists who work to help people with coping difficulties are *clinical psychologists*. The types of problems they help with are extremely varied, but all are considered to be psychological disorders. In addition, many non-psychologists, such as *psychiatrists*, *social workers*, *psychiatric nurses*, and *school counselors*, also work with people who have these sorts of problems. In order to guide practitioners with these varied types of problems, which call for varied types of assistance, extensive classification systems have been created. Because different professions use these classification systems, they have developed independently of any single helping profession.

In the United States, the principal classification system used is the *Diagnostic and Statistical Manual of Mental Disorders* (currently in its Fourth Edition), otherwise known as the *DSM*. Interestingly enough, the DSM is written and published by psychiatrists, not psychologists, so it doesn't always reflect the views of psychologists on various issues. Nevertheless, the DSM is the cornerstone of the medical model, and the medical model is the principal way in which clinical psychologists determine what sort of help people with different sorts of problems need.

Key Point: Assumptions of the Medical Model

While it may seem obvious that psychological problems should be classified in the same way as medical problems, this is only because we have become accustomed to doing things this way. From a scientific perspective, the *medical model* involves a number of important assumptions about what psychological disorders are and how they should be treated. It is important to remember that not all of these assumptions are true in all cases all of the time.

Many psychological problems are most likely genuine diseases, no different from other medical problems, except that the symptoms are mostly mental, rather than physical. Some psychological problems are merely manifestations or side-effects of physical problems that are better treated by medical means. Other psychological problems involve behaviors that are more problematic for society in general than they are for the person with the problem. As such, the legal system, in particular the criminal justice system, often becomes involved.

In addition, many of the psychological problems that people face in coping with the world may be caused by healthy minds having learned to cope with unhealthy circumstances. Bad behavior patterns can develop when a person is forced to adjust to bad situations. Those bad behaviors may persist even when the person gets out of the bad situation. In these cases, the reason that the psychological problems can be classified according to their behavioral similarities is not because they reflect the same problem within the individual, but because these people faced the same sorts of problematic situations in society.

In medicine, we classify medical problems both in order to determine what currently available treatment is best and also to help find the cause of the problem in order to work towards a cure. In psychology, it is often useful to determine the best treatment to use. However, the cause of the problem (in the world) may not be of much help in the cure (of the individual).

HISTORY

Psychological problems have been known throughout history, with the more dramatic problems such as schizophrenia and mental retardation being identified earliest. The causes of these problems have been subject of much debate, which continues to this day. The ancient Greeks thought the causes were due to fluid imbalances within the body, just like other medical problems. In medieval Europe, schizophrenia and other problems resulting in bizarre behavior were thought to be the result of demonic possession and witchcraft. The modern medical model, proposed in the eighteenth century, had the distinct advantage of lessening the cruelty to which those with psychological disorders were often treated in European countries and their colonies.

In the twentieth century, advances in diagnosis helped pave the way for very rapid advances in treatment and even cures for many medical problems. Progress with psychological disorders during this period was slower. Diagnostic criteria played key roles in determining that certain psychological disorders were, in fact, physical maladies caused by infection or genetic abnormalities. However, the medical approach thus far has failed to yield solid diagnostic criteria, a thorough understanding of the causes, or effective treatments for the majority of psychological disorders.

CURRENT APPROACHES

There are two current approaches to understanding psychological disorders: the medical model and the biopsychosocial model. Currently, researchers and practitioners using the medical model are focused on biological causes and pharmacological treatments. Recent advances in genetics and brain scanning have made it possible to look for biological contributors to various psychological disorders. Every tiny hint that any psychological disorder may have a genetic or brain-based cause is pursued. Of course, this diminishes the focus on nonbiological contributing causes. Recent advances in pharmacology have resulted in the development of a large number of drugs with some beneficial effects for a small number of psychological disorders, particularly the mood disorders. More and more drugs with similar effects are being developed, and these drugs are being tested on other disorders. Often, persons with multiple disorders are treated for the disorder for which a drug exists and the other disorders are neglected.

The biopsychosocial model is the current challenger to the medical model. It is often laid out in very theoretical terms. An analogy may help here. Think of a car radio that is producing a lot of static and noise. What might be wrong? The tuner could be between stations. The signal could be blocked by an overpass, bad weather, or a passing truck. The radio might have been badly designed, or poorly constructed, or it could just be broken. Now, suppose we know nothing about radios or electricity

Variant View: The Mystery of "Underlying Depression"

Currently, there are a large number of drugs available for treating different types of depression. These antidepressant drugs have their problems, not the least of which is that many are effective only so long as they are taken. Regardless, these drugs have converted depression from one of the many almost untreatable psychological disorders into one of the most treatable. Depression co-occurs with many other disorders, however. It is also no easy matter to distinguish depression from sadness or grief, much less to differentiate between the various types of depression.

Recently, proponents of the medical model have argued that people with various conditions, including substance abuse, anxiety disorders, childhood behavioral disorders, and even simple old age, may have depression, but the symptoms are masked by the other condition. In some cases, closer examination or improvement in the other condition may reveal symptoms of depression. However, there is a growing view that some of these other disorders always or almost always co-occur with depression, whether or not symptoms of depression are detected.

One possible reason for this view is that some individuals with these other conditions show behavioral improvements when given antidepressant drugs. Of course, another equally plausible explanation is that the daily difficulties and stresses of having these other conditions help trigger depression, or merely make the person very sad and angry. The real problem is that antidepressants are prescribed and depression is diagnosed in the absence of sufficient symptoms. A central basis for the DSM system is to specify behavioral diagnostic criteria for medical conditions for which nonbehavioral symptoms are unavailable. Basing a diagnosis upon the efficacy of medications for which prescribing conditions have not been met and whose underlying mechanism and relationship to the diagnostic category in question are not well-known, stands this logic on its head.

This is not to say that there is some grand conspiracy of drug companies and doctors to over-diagnose depression in order to sell more pills. Healthcare providers are motivated by a desire to help and are frustrated when cures are not available. New treatments provide them with hope and are often overused initially. The practitioner's desire to cure must be balanced against the scientific need to research and develop safe and effective treatments over the long term.

and precious little about cars. How would we go about finding out what is wrong and/or doing something about it? To eliminate the possibility that we are between stations, we can adjust the tuning. We can look around to see if there are any tunnels or other obstructions or to check the weather. Once we have eliminated the things we cannot easily examine or control, we can begin to speculate about things within the radio itself.

The biopsychosocial model emphasizes the examination of the biological, psychological, and social contributions to psychological disorders. Instead of focusing on just the biological (the "bio"), this model also looks toward the individual's psychological dynamic (the "psycho"), and the larger environment in which the individual functions (the "social"). The heart of the biopsychosocial model is that not only are there different sorts of causes for psychological disorders, but the causes can interact. A badly designed and/or poorly constructed radio is more likely to lose the signal in bad weather, for example. As we will see in the next section, the medical model leads to a focus on biological causes to the exclusion of other causes.

The DSM Classification System

The importance of the DSM in the treatment of psychological disorders in the United States cannot be understated. While developed by the American Psychiatric Association, it is the cornerstone for diagnosis for all sorts of practitioners. Over the decades, the psychiatric community has worked to align the DSM both to the theoretical orientations and practical necessities of both the therapeutic and research communities. There has also been a lot of politics. In this section, we will discuss two issues that arise in the development of a classification system: the question of identifying what is and is not a psychological disorder and the problems that arise from applying the medical notion of diagnosis to psychological problems.

WHAT IS A PSYCHOLOGICAL DISORDER?

The first issue faced by anyone developing a classification system is what to include. What is a psychological disorder? In no small part due to the fact that a medical code number is required for insurance reimbursement and other regulatory and legal matters, the DSM has tended to err on the side of inclusion. Any problem that might be construed as being psychological, even if only in part, is very likely to be found in the DSM.

The authors and editors of the DSM freely admit that there is no possibility of a strict definition of what is and is not a psychological disorder. They provide a definition as guidance that specifies the obvious things: a pattern of behavior that is disruptive, potentially harmful, deviant, or makes the patient unhappy, with the specific exclusion that the behavior not be a culturally sanctioned reaction to some event. Of course, by this definition, mountain climbing, ordinarily considered a sport, is a psychological disorder. (One imagines that, over a century ago, when Europeans first arrived at the Himalayas intent on climbing them, the locals justifiably

Tips on Terms: Major Categories of the DSM

- **Disorders usually first diagnosed in infancy, childhood, or adolescence** Includes mental retardation, autism, attention-deficit and behavioral disorders, and learning disabilities.
- **Delirium, dementia, and amnesic and other cognitive disorders** Includes Alzheimer's, amnesia and cognitive impairments due to organic causes, and senile dementia.
- **Mental disorders due to a general medical condition**
- **Substance-related disorders** Includes dependence, abuse, intoxication, and withdrawal subdivided by various drugs and substances, including both legal and illegal drugs.
- **Schizophrenia and other psychotic disorders** Includes schizophrenia and delusional disorder.
- **Mood disorders** Includes depressive and bipolar disorders.
- **Anxiety disorders** Includes panic disorders, phobias, obsessive-compulsive disorder, and post-traumatic stress disorder.
- **Somatoform disorders** Includes somatization, conversion, and pain disorders, and hypochondriasis.
- **Factitious disorders** Includes faking symptoms specifically to take on a sick role.
- **Dissociative disorders** Includes amnesia, fugue, and dissociative identity disorder.
- **Sexual and gender identity disorders** Includes dysfunctions, deviancies, and identity problems, etc.
- **Eating disorders** Includes anorexia and bulimia.
- **Sleep disorders** Includes insomnia, narcolepsy, sleep apnea, and sleepwalking.
- **Impulse-control disorders not elsewhere classified** Includes kleptomania, pyromania, and pathological gambling.
- **Adjustment disorders** Includes reactions to stressful life events or circumstances.
- **Personality disorders** Includes perduring, pervasive, and inflexible deviation outside the normal range of personality variations.
- **Other conditions that may be a focus of clinical attention** Includes problems with medication, relationships, abuse and neglect, and problems with specific life circumstances.

thought them mad.) Mountain climbing is not included in the DSM, but exhibitionism (technically, exposing one's genitalia to a stranger) is. In the United States, there are few culturally sanctioned outlets for males with exhibitionistic tendencies. Males who act on their exhibitionistic tendencies tend to get thrown in jail. Females, on the other hand, are more likely to find employment as nude models, exotic dancers, or in the adult entertainment industry. These women would not fit the specific DSM criteria for exhibitionism because one criterion requires that the behavior cause "interpersonal difficulty."

Despite the enormous breadth of coverage in the DSM, there is a consistent theme in the understanding of psychological disorders. Throughout the book, the phrase "distress or impairment" appears in the criteria for various psychological disorders. In other words, an ongoing pattern of behavior is evidence of a psychological disorder if it causes the patient to complain (distress) or other people to complain about the patient (impairment).

THE PROBLEM OF DIAGNOSIS

Once the overall scope of a classification system is determined, the next problem is to determine whether a patient has a psychological disorder and, if so, which one. It is in this key area that the medical model shows both its greatest strengths and greatest weaknesses. By attending to specific patterns of behavior found within the clinical population, psychological disorders with specific physical root causes were distinguished from other psychological disorders and effective treatments found. Concern for subjective and variable diagnostic practices atypical of other medical professions led psychiatrists to include specific diagnostic criteria beginning with the Third Edition of the DSM (published in 1980). This practice has been expanded with the current edition, with a specific eye toward both making use of and promoting further research into *differential diagnosis.*

The great weakness of the medical model with respect to diagnosis was pointed out by Thomas Szasz (1961). Szasz, within the context of a larger critique of the medical model in which he challenged the very notion of mental illness, argued that

Tips on Terms: Differential Diagnosis

Differential diagnosis is a medical term for distinguishing between two or more disease conditions, any of which might be present given the patient's initial presentation of symptoms. When treatments are available, differential diagnosis is invaluable in ensuring that the patient receives the right treatment. When treatments are not available, differential diagnosis can assist researchers in finding the different causes of specific diseases.

since the distinction between psychologically disordered and not disordered was based on a relationship between the person's behavior and the social environment, examining just the person and his behavior might prove insufficient for diagnosis. A critical difference between diagnostic techniques for psychological disorders and for other medical problems is the availability of corroboration using nonbehavioral tests, such as blood tests, tissue pathology, and so on. The last two editions of the DSM have taken important steps forward in providing checklists of behavioral symptoms and requiring at least two symptoms on each list be present for diagnosis. But this sort of corroboration is weaker than detecting syphilitic bacteria in a test tube. Not only are behavioral symptoms more subjective in the sense of requiring clinical judgment and expertise, but also the mere fact that *all* of the DSM criteria for one disorder can be behavioral means that there is often no way to double-check the subjective judgments, except with more subjective judgments.

The weaknesses pointed out by Szasz were further illustrated empirically by Rosenhan (1971). Rosenhan sent fake patients to various mental hospitals with a single fake symptom that could be interpreted as a novel hallucination. Eight out of eight were admitted and diagnosed with serious disorders. Further examinations failed to detect the fraud. Rosenhan was careful to point out that his study did not challenge the medical model, but only the possibility of errors in diagnosis. However, the Rosenhan study was influential in the radical changes in commitment laws that followed in the 1970s. More recently, some psychiatrists have challenged Rosenhan's study. They have suggested that the likelihood of the success of deliberate fraud and malingering is not a good measure of actual errors in diagnosing real patients. They point out that a person might secrete blood in her mouth and spit it out to fake a stomach ulcer or claim numbness and the inability to move in order to fake a stroke. These examples are specious. Stomach ulcers often cause abdominal pain and tenderness. The inflammation can be observed endoscopically. While the patterns of numbness and paralysis from stroke vary widely, faking a realistic pattern is not an easy matter. Few nonpsychological disorders could be faked effectively by pretending to have just one symptom. Szasz was correct. The basic process of diagnosing other medical disorders is less prone to error than is psychiatric diagnosis.

Of course, once upon a time, most physical ailments were equally difficult to diagnose. Improvements in medical technology promise that more psychological disorders, at least those with significant biological components, will be diagnosable with nonbehavioral measures. Checklists in the DSM, not available at the time of the Rosenhan study, lessen the likelihood that faking a single symptom will result in misdiagnosis. The mere existence of the controversy, along with changes in civil liabilities law, means that clinical practitioners are more cautious.

Specific Disorders

At an introductory level such as this, it is impossible to consider more than a tiny fraction of the psychological disorders classified in the DSM. The question of why to discuss a disorder is ultimately more interesting than which disorders are chosen. There are at least five good reasons for discussing a psychological disorder at the introductory level:

- **Historical/classificatory** Some psychological disorders are important because of the role they played in the development of our current understanding of and treatments for psychological disorders.

- **Prevalence/incidence** If many people have a disorder, it is important.

- **Controversy** Controversial disorders are important because issues are unsettled.

- **Social impact** Even uncommon disorders may be important if they impact society disproportionately to their prevalence/incidence.

- **Treatable** The availability of effective treatments for a particular disorder affect diagnosis, treatment, and even classificatory systems of disorders in general.

HISTORICAL/CLASSIFICATORY

Some psychological disorders are important because of the role they played in the development of our current understanding of and treatments for psychological disorders. In light of their historical importance, we will look at *schizophrenia* and two of the somatoform disorders, *conversion disorder* and *somatization*.

Handy Hints: First-year Medical Students' Disease

A common phenomenon among psychology students is the tendency to succumb to "first-year medical students' disease." This is the tendency to learn about a new disease (or, in this case, a psychological disorder), recognize that you experience some of the symptoms, and incorrectly conclude you have the disorder. While this "disease" is not serious, it can cause the student undue stress. Particularly with psychological disorders, it is important to realize that many of the symptoms are extreme, debilitating, and intense versions of thoughts and behaviors that healthy people experience every day. At various moments, we can be anxious, depressed, withdrawn, obsessive/compulsive, suspicious, deluded, or anti-social. Unless the overall pattern of these experiences meets the criteria of a psychological disorder, they are considered perfectly normal.

Handy Hints: Split Minds and Multiple Personalities

Schizophrenia is also one of the most misunderstood disorders by those who have not studied psychology. The most common mistake is confusing *schizophrenia* with *multiple personality disorder* (now referred to as *dissociative identity disorder*), a completely different disorder with different symptoms, causes, and treatment. This confusion stems from the fact that the word, *schizophrenia,* literally translates as "split mind," referring to the disintegration of thought processes seen in the disorder. Unfortunately, fiction writers, news reporters, and others in the popular media have confused "split mind" with "split personality."

Schizophrenia

Historically, the psychological disorders with the most dramatic symptoms were the first to be recognized and named. The stereotype of a mentally disturbed person is one who rants, talks nonsense, believes in impossible things, and sees and hears things and people that are not there. While these symptoms can appear as part of various conditions, they are most characteristic of schizophrenia. Schizophrenia is the prototypical psychological disorder and where our understanding of psychological disorders begins.

Schizophrenia is a general term for a class of psychological disorders that involve disturbances in thoughts, emotions, perception, and/or language. The symptoms of schizophrenia can affect a wide range of perceptions, emotions, behaviors, and cognitions. Importantly, one schizophrenic patient usually will not exhibit all symptoms. Indeed, the classification of subtypes is based on the prevalence of a specific type of symptom:

- **Perceptual symptoms**   Perceptual symptoms are those symptoms that affect how we perceive reality.

  - *Hallucinations* are sensory experiences that are not connected to any sensory stimulation. Hallucinations are most commonly auditory. Patients report hearing voices that make insulting comments or give instructions, but may also be complete nonsense.

  - *Delusions* are unfounded or false beliefs that are clearly out of touch with reality, affecting how a patient copes with the world. There are four types of delusions: Delusions of *persecution* involve the belief that people are out to get you. Delusions of *grandeur* involve the belief that you are especially important or that you are a famous person. Delusions of *reference* involve the belief that ordinary events are actually messages intended only for you. Finally, delusions of *control* involve the belief that your feelings, behaviors, or thoughts are controlled by others.

Note that all of these delusions have a common thread of defending against the notion that the patient is unimportant.

- **Affective symptoms**   Affective symptoms influence how the patient experiences and displays emotions. In most instances, patients do not express much emotion, either in facial expressions or voice inflections. This is known as *flat affect*. When a patient does show emotion, it is often inappropriate to the situation in intensity, manner, and type.

- **Behavioral symptoms**   Behavioral symptoms affect the actions and movements of the patient. An extreme case is *catatonia*, where a patient may be completely motionless for a period of time before going into a state of hyperactivity. A more common behavioral symptom involves *stereotypies*, senseless repetitive actions, such as rocking back and forth, or rubbing one's arm.

- **Cognitive symptoms**   Cognitive symptoms indicate the deterioration in the thought and language processes of the patient. The thought processes, and the associated language they produce, are often disorganized. Patients often make *loose associations* between concepts, jumping from one idea to the next, often within the same sentence.

As mentioned earlier, schizophrenia is not a single disorder, but a cluster of disorders that have some common sets of symptoms. Depending on which symptom or set of symptoms is predominant, a patient may be classified as having one of four subtypes of schizophrenia:

- *Paranoid* schizophrenia is dominated by perceptual symptoms.
- *Catatonic* schizophrenia is dominated by behavioral symptoms.
- *Disorganized* schizophrenia is dominated by cognitive symptoms.
- *Undifferentiated* schizophrenia is diagnosed if no single set of symptoms predominates.

Somatoform Disorders

Somatoform disorders, particularly somatization disorder and conversion disorder, are important to the history of psychology because, while uncommon today, they are the types of disorders that first attracted Freud to the study of the psyche. These physical ailments without physical causes made Freud look elsewhere for the true cause. Peculiarities in the patterns of these symptoms led Freud to hypothesize an organization underlying mental illnesses. While much of this hypothesis has proven false, the essential notion of a "logic" to psychological diseases continues to be a central part of treatment for many disorders. Indeed, when a disorder, such as schizophrenia or autism, is shown to lack an internal logic, this fact alone is important evidence of an as yet-to-be-discovered physical cause.

People diagnosed with somatoform disorders suffer a wide variety of physical ailments. What makes this a psychological disorder is that the underlying cause is psychological, rather than organic. Importantly, a patient with a somatoform disorder is not *malingering*, that is, faking symptoms in order to get attention or escape life's unpleasant duties. The physical symptoms are real, but the cause is psychological.

Conversion disorder is marked by the loss of physical function or sensation, usually in a single organ system, with no physical underlying cause. Common symptoms involve the loss of vision (previously known as *hysterical blindness*), partial or complete paralysis, or the loss of feeling in a limb. Often, the psychological underpinnings are relatively easy to discover, as the medical symptoms experienced by the patient are not consistent with what medical science knows about anatomy or the nervous system. In some cases, the underlying psychological cause is directly tied to the type of symptoms the patient experiences. For example, a case of blindness may be tied to witnessing an emotionally disturbing event.

Somatization disorder, once known as hysteria, is the psychological disorder that first attracted Freud's attention. It is a pervasive disorder of long duration, involving at least gastrointestinal symptoms, several types of pain, at least one sexual symptom, and at least one neurological conversion symptom other than pain. Both the frequency of the disorder and the pattern of symptoms vary across cultures and over time within one culture.

As medical science has progressed, far more subtle medical conditions can be diagnosed. Many diseases not known in Freud's time, particularly the autoimmune diseases such as multiple scherosis and lupus, can also present with a wide variety of vague symptoms. Before a somatoform disorder is diagnosed, the physician must be careful to exclude any possible organic cause of the symptoms.

PREVALENCE/INCIDENCE

If many people have a disorder, it is automatically important to society. Here, we will look at two common disorders, *depression* and *obsessive-compulsive disorder*.

Depression

Mood disorders are characterized by emotional disturbances that can disrupt physical, social, or cognitive functioning. Depression is the most common form of mood disorder. During a depressive state, a person will experience severe sadness and hopelessness. These emotional disturbances can be episodic and alternate with periods of normal emotional states. Thus, people with mood disorders can often function normally until they experience an episode of severe mood change.

Everyone feels depressed at some point in his life. What distinguishes these normal feelings from a mood disorder is the cause of the mood and the severity and duration of the symptoms. If a depression stems from a specific event, such as the death of a loved one, it is defined as a *reactive* depression and not a mood disorder. However, if the emotional disturbance is not tied to a specific event, lasts for more than two weeks, and disrupts daily functioning, it is classified as a mood disorder.

Major depressive disorder is the most severe form of depression. It is characterized by a persistent depressed mood that does not stem from a specific event and is accompanied by a loss of interest in pleasurable activities. In addition to these primary symptoms, patients will also experience at least four of the following: physical symptoms including unintentional weight loss, fatigue, and extreme physical agitation or extreme sluggishness; sleep disturbances resulting in either the inability to sleep (insomnia) or in sleeping too much (hypersomnia); disruptions in cognitive functioning, such as feelings of worthlessness or guilt; diminished ability to concentrate; and recurrent thoughts of suicide.

Obsessive-compulsive Disorder

Obsessive-compulsive disorder (OCD) is characterized by irrational obsessions and compulsions that interfere with a person's daily functioning. Obsessions are persistent, unwanted thoughts or ideas. Compulsions are repetitive behaviors usually enacted to address the concerns expressed by the obsessive thoughts. For example, a person may be obsessed with the idea that there are deadly germs living on her doorknob. In response to this obsession, the person is compelled to clean the doorknob over and over again, even when it is already clean. There is nothing wrong with a clean doorknob. On the other hand, obsessing about germs and compulsively cleaning the doorknob 300 times a day most definitely disrupts one's functioning.

CONTROVERSY

Controversies, either among professionals or in the public sphere, make a disorder important. Here, we will look at *post-traumatic stress disorder*, *dissociative identity disorder*, and the general category of *personality disorders*.

Post-traumatic Stress Disorder

Post-traumatic stress disorder (PTSD) occurs as a consequence of a traumatic event, such as war, abuse, or a disaster. Three conditions must be met in order to diagnose PTSD: First, a person must have experienced an event that involves actual or threatened serious injury or death. Second, a person must have responded to the situation with helplessness or fear. Finally, the person experiences three sets of symptoms as a result of the experience: persistently re-experiencing the traumatic event, either in dreams or flashbacks; avoidance of anything associated with the traumatic event; and heightened arousal, which can lead a person to startle easily and have difficulty sleeping.

While PTSD itself is not controversial, controversies have tended to surround it. PTSD was first discovered in combat soldiers, but the symptoms appeared only after the soldier was safely removed from the combat. This fact caused some to falsely suspect these soldiers of malingering to avoid return to active duty. The diagnostic criteria for PTSD now make this misdiagnosis very unlikely. The delay in the onset of symptoms may only be a matter of hours, but can also be a matter of months or years. Recently, some psychotherapists have begun to use hypnosis and related techniques to establish memories of hitherto unknown childhood traumas in patients who do not initially present with the full pattern of PTSD symptoms. These therapists have proposed a theory of *recovered memory syndrome*, which is very controversial. The idea is that many different types of psychological disorders may have long-forgotten childhood traumatic events as their cause. The absence of some PTSD symptoms and memories of the event prior to therapy, and the difficulties in establishing the actual occurrence of those events by other means, present very real problems. Research indicates that so-called recovered memories may actually be induced by the hypnosis or other techniques used to detect them.

Dissociative Identity Disorder (formerly Multiple Personality Disorder)

Dissociative disorders are rare disorders in which a person experiences loss of memory. Conscious awareness becomes separated, or dissociated, from previous memories. Dissociative disorder is only diagnosed when the memory loss does not stem from physical injury. It is also important to distinguish this from simple forgetting. Personal memories (one's name or life history), and not general facts (the alphabet or how to drive a car), are the focus of memory loss.

Dissociative disorders are thought to be very rare, although the prevalence of *dissociative identity disorder* (DID) is controversial. DID occurs when a person's personality fractures into two or more distinct identities. This purportedly stems from severe psychological stress, typically repeated sexual and/or physical abuse, experienced during childhood. It is believed that the additional identities are a coping mechanism created to protect the victim. Thus, the alternate personalities can cope with situations that the individual (or "original" personality) cannot. When we examine the characteristics of the alternate personalities, we see that each identity has its own name, gender, memories, personality traits, and physical mannerisms, and can be completely different from the original identity. Thus, if the person is quiet and shy, an alternate personality may be outgoing and flamboyant. This personality may emerge when the situation calls for social interaction. Another personality may be older and emerge when the individual needs to exert authority.

The controversy surrounding DID is related to that of recovered memory syndrome discussed previously in terms of PTSD. Prior to 1960, fewer than ten people were found to have DID. Beginning in the 1980s, some psychotherapists, believing that childhood trauma was the cause of DID, began to use hypnosis and other techniques on a large scale to recover memories of these events. Suddenly, there was a huge, unprecedented increase in the number of reported cases. Cases are now reported in the tens of thousands each year. These cases are found almost exclusively in the United States and are entirely absent in places where only few psychotherapists use recovery techniques. Even in countries where DID is found, it is found almost exclusively by a small number of therapists. There is little in the way of explanation as to why this unprecedented increase in cases in such a limited number of societies should have occurred with such enormous speed and be detectable by only a small number of therapists.

Personality Disorders

Personality disorders are something of a grab-bag of less serious conditions. What personality disorders have in common is that each represents an extreme of tendencies found normally in the population. When the tendency is not only extreme, but also pervasive and inflexible, a personality disorder may be diagnosed. A personality disorder is diagnosed only when the trend begins no later than early adulthood and persists for some years without remission. It must either disrupt functioning or be painful to the patient.

Some personality theorists object to the notion that personality disorders are disorders on the basis that what constitutes an extreme personality trend will vary from society to society, and from societal role to societal role. Those with family members, friends, and loved ones who have personality disorders may disagree. Personality disorders, even when they do not cause distress to the patient, always limit the patient's social relations in difficult and often painful and disruptive ways. Unlike other personality problems, where a personality trend might involve painful or disruptive behaviors, personality disorders resist treatment.

Here are the most commonly diagnosed personality disorders:

- Cognitive trends
  - **Paranoid**   Characterized by suspicion of others
  - **Schizoid**   Characterized by social isolation, emotional detachment, and limited emotional expression
  - **Schizotypal**   Characterized by odd beliefs, eccentricities, unusual speech, and social limitations

- Mood trends
  - **Anti-social**   Characterized by impulsivity, deceit, disregard for the feelings and rights of others, violations of rules and laws, and violence.
  - **Borderline**   Characterized by highly erratic and changeable interactions with others, behavior, and attitudes, coupled with impulsivity, cruelty, and fear of abandonment.
  - **Histrionic**   Characterized by extreme emotionality and need for attention.
  - **Narcissistic**   Characterized by a sense of entitlement, neglect of others, boastfulness, need for praise, and fragile self-esteem.
- Anxious trends
  - **Avoidant**   Characterized by social isolation, shyness, secretiveness, and fear of criticism.
  - **Dependent**   Characterized by indecisiveness, need for close relationships, lack of anger, and lack of self-confidence.
  - **Obssessive-compulsive**   Characterized by perfectionism, rigidity, and neglect for the opinions of others in the absence of the true obsessions and compulsions found in obsessive-compulsive disorder.

SOCIAL IMPACT

Even uncommon disorders may be important if they impact society. Just one individual whose psychological problems turn him into a serial killer is one too many. A disorder that affects very few people, but comprehensively disables them so that they require very expensive, lifetime care, is important. Here we will look at *substance abuse* and the general category of *developmental disorders*.

Substance Abuse

Substance abuse is one of the easiest problems to identify and one of the hardest to define as a psychological disorder. The use of substances that have psychological effects is easily determined, but what constitutes abuse? In US society, substance abuse occurs when either the substance or its use is disapproved of. This societal component complicates matters and creates controversy in that medical, legal, political, sociological, and even religious issues are involved. The DSM attempts to avoid these problems by dividing substance-related disorders into those problems directly caused by substance use on one hand and dependence and abuse on the other.

Even when dealing with symptoms directly caused by intake of a substance, definitional problems arise. When should intoxication be considered a psychological disorder? The DSM deals with this by recourse to the standard requirements of distress or impairment. The DSM criteria for substance dependence also include the distress or impairment requirement, but even this is not sufficient to address the societal issues. The traditional model of physiological dependence, based on early research on opiates, had to be revised to deal with dependence on other types of drugs, and this model was heavily influenced by political considerations of what drugs were disapproved of and/or illegal. Cocaine, in particular, presented problems. Societal disapproval of cocaine throughout the twentieth century rose and fell in cycles of about 30 years duration. Cocaine dependence operates very differently from dependence on opiates, alcohol, barbiturates, and other drugs. The political requirement that cocaine addiction be treated as other addictions led to an expansion and complexification of the criteria for substance dependence.

Even with these other drug-related problems separated out, this leaves a core of problems classified in the DSM as substance abuse. The criteria for substance abuse are very broad. Any single type of distress or impairment (including legal problems) due to ongoing substance use, where substance dependence is *not* found, is classified as substance abuse. This permits intervention by medical professionals, clinicians, and healthcare providers.

Developmental Disorders

Developmental disorders arise during the course of psychological development in childhood or adolescence. Individuals with developmental disorders may have enormous difficulties functioning without assistance. As such, the problems of those with developmental disorders often become problems for their families and for society as a whole. We will discuss two sorts of developmental disorders: *mental retardation* and *autism*.

Mental retardation occurs when the normal cognitive and intellectual development of the child slows, ceases, or reverses. The defining characteristic and net result in adulthood is a person who lacks the general cognitive capacities (what we think of as "intelligence") to function effectively in society without assistance. Because intelligence itself is not well understood and is measured on a continuum, it is unclear how limited one's intelligence must be before one can be considered to have a psychological disorder. Everyone would probably manage life a bit better if he or she were a bit smarter. Many cognitively impaired persons have other skills that vastly exceed those of people with normal intelligence. The cutoff scores that differentiate the mentally retarded from the rest of us are necessarily arbitrary. It is the general functional impairment that distinguishes the mentally retarded individual.

Autism is one of the Pervasive Developmental Disorders (PDDs), also known as Autistic Spectrum Disorders. Like mental retardation, autism and the other PDDs appear in childhood and involve impairments to various capacities. Unlike mental retardation, the impairments are not specifically cognitive. While some persons with autistic disorders may have severe cognitive impairments, others have normal or even above normal IQs. The central feature common to the autistic spectrum disorders is impairment to social functioning and communication. In severe cases, this can result in the total absence of language (mutism).

Developmental disorders of all types are most likely physical abnormalities of the brain, although current technology has not yet determined the specific abnormalities in most cases. These abnormalities may be caused by injury or by genetic factors or by a combination of both. For example, an extra chromosome 21 produces Down Syndrome, a developmental disorder that causes both cognitive impairments and physical disabilities. For the most part, developmental disorders are not "mental" illness in that there is almost certainly a physical cause, even if it has not yet been discovered. Counseling and purely behavioral treatments can have important beneficial effects in helping the developmentally disabled function in society, but they do so by enabling the patient to use the capacities he has more effectively, rather than by correcting the underlying limitations.

TREATABLE

Effective treatments for psychological disorders are in short supply. In our discussion of depression, we saw how, in this environment, effective treatments can have negative impacts on clinical practice in general. However, a disorder for which effective treatments are available is important for the lessons it can teach us about treating other psychological disorders. Here we will look at two very treatable disorders, *bipolar disorder* and *phobias*.

Bipolar Disorder

Bipolar disorder was the first of the mood disorders to be treated effectively using drugs. As such, it has become a paradigm for pharmacological intervention in psychological disorders. Bipolar disorder is characterized by severe mood swings from depression to mania. "Bipolar" refers to the cyclical movement between the two poles of the emotional continuum of depression and mania. The depressive episodes are very similar to those found in major depressive disorder. The manic episodes are characterized by extreme elation, feelings of euphoria, and a sense of invulnerability. During a manic phase, a person can feel very sociable and confident, tend to engage in impulsive and risky behavior, and experience thoughts that jump from one idea to the next. It is important to note that, while a manic episode may not sound as negative as the depressive state, the behavioral effects can be just as devastating. While a mild manic episode can result in confidence and increased productivity, more severe episodes can lead to behavior that is not only unproductive, but very risky, illegal, and sometimes fatal.

Phobias

Phobias were the first psychological disorders to be treated effectively using behavioral techniques. Understanding phobias can help us understand when and where behavior therapy can be useful, either as a primary or adjunct treatment. Phobias are intense, irrational fears of specific objects, activities, or situations. Most of us experience a mild phobia of spiders, heights, and the like. However, those diagnosed with a phobia are those whose fear is so intense and debilitating that it interferes with their daily life. Their anxiety often escalates into panic attacks and avoidance behavior. Panic attacks involve feelings of intense anxiety and dread that are not justified by the situation. These psychological symptoms are often accompanied by physical symptoms, such as chest pains, choking, difficulty breathing, or fainting. Avoidance behavior involves going out of one's way to avoid situations that trigger phobic attacks. Avoidance behavior can expand to the point of disrupting the patient's life by preventing him or her from engaging in normal activities, even to the point of not leaving home.

What distinguishes phobias is that the reaction is directed toward a specific *fear object*. The DSM classifies specific phobias as *animal type, natural environment type* (including fear of heights, storms, caves, and so on), *blood-injection-injury type, situational type* (including fear of airplanes, elevators, and so on), and *other type* (including fear of choking, clowns, and so on). The older classification system gives a better idea of the breadth of fear objects found in the clinic. It is also more fun.

Here are some of the more common phobias:

- **Acrophobia**   Fear of heights
- **Claustrophobia**   Fear of enclosed spaces
- **Agoraphobia**   Fear of open spaces or travel
- **Aelurophobia**   Fear of cats
- **Algophobia**   Fear of pain
- **Astraphobia**   Fear of storms
- **Hematophobia**   Fear of blood
- **Nyctophobia**   Fear of darkness
- **Apiphobia**   Fear of bees

- **Enetophobia** Fear of needles
- **Herpetophobia** Fear of reptiles
- **Iatrophobia** Fear of doctors

Summary

As we have seen, there are a huge number of psychological disorders, classified and sub-classified according to an intricate system that changes not only with scientific advances, but also with the cultural assumptions about what constitutes a mental disorder. In this book, we only have enough space to deal with a tiny fraction of these problems, and none of those in depth. There are at least two reasons for such an elaborate classification system. The first is purely scientific: Psychological disorders vary widely. The DSM classification system works to represent this complexity accurately. The second reason is practical: Different disorders require different sorts of treatments. We will discuss treatments in Chapter 8.

Quiz

1. Which of the following helping professionals are clinical psychologists?

   (a) Psychiatrists

   (b) Social workers

   (c) School counselors

   (d) None of the above

2. What are the two basic approaches to thinking about psychological disorders?

   (a) The DSM and the biopsychosocial model

   (b) The medical model and the biopsychosocial model

   (c) The medical model and the DSM

   (d) The psychiatric approach and the psychological approach

3. Which of the following is NOT considered to be a psychological disorder?

   (a) Spousal abuse

   (b) Homosexuality

(c) Playing sick

(d) None of the above are considered psychological disorders.

4. The DSM uses _____ to determine whether symptoms are classified as a psychological disorder.

(a) Distress and impairment of the patient

(b) The organic cause of the symptoms

(c) A direct link to psychological stressors

(d) Depression

5. Which of the following is a recent improvement in the DSM, beginning with the Third Edition?

(a) Checklists for diagnosis

(b) Specification of nonbehavioral symptoms for organic problems

(c) Specific rules for differential diagnosis

(d) All of the above

6. _____ schizophrenia is characterized by predominantly cognitive symptoms.

(a) Paranoid

(b) Catatonic

(c) Disorganized

(d) Undifferentiated

7. Which psychological disorder is considered important because of its high incidence in the population?

(a) Schizophrenia

(b) Depression

(c) Autism

(d) Bipolar disorder

8. _____ has caused controversy in the diagnosis of post-traumatic stress disorder and dissociative identity disorder.

(a) The biopsychosocial model

(b) The medical model

(c) Recovered memory syndrome

(d) All of the above

9. According to the DSM, substance _____ is considered to be separate from substance abuse.

   (a) Dependence

   (b) Withdrawal

   (c) Intoxication

   (d) All of the above

10. _____ and _____ have proven to be successful in treating some psychological disorders.

   (a) Counseling, drug treatments

   (b) Behavioral therapy, counseling

   (c) Drug treatments, behavioral therapy

   (d) Drug treatments, surgery

CHAPTER 8

Psychotherapy

We discussed the various types of psychological disorders in the previous chapter. Here, we'll discuss the various types of treatment for those disorders.

Overview

When people think about psychology, they most often think about practitioners, clinicians who seek to treat mental disorders. As discussed in previous chapters, there are many other types of psychologists and many other parts to psychology. In this chapter, we will take a look at clinical treatments and how they relate to the science of psychology more generally.

HISTORY

Attempts to treat emotional and mental problems date back to ancient times. With the advent of Christianity in the West, these efforts became closely tied to a theory of the causes of mental disease based on possession by demons. With the

Renaissance came reform. As emotional and mental problems came to be seen as diseases, treatments based on the medical model began, returning treatment to its classical roots. Each type of therapy was developed with respect not only to the inventors' views of the human mind, but also to their ideas about medicine.

In the late nineteenth century, Freud, trained as a physician, came to recognize that disorders of the mind resisted physically based therapies and even physically based explanations. Although he believed that the mind would eventually be explained in bodily terms, Freud sought to develop mental analogues to the anatomy and physiology of the body for understanding the mind. He then sought mental analogues to pharmaceuticals and the scalpel. After rejecting hypnosis, Freud decided that highly structured conversation with the patient was the way to reach into and alter the mind. This was the first *insight therapy*, the first type of *talk therapy*.

With the rise of the behaviorism in the 1920s, critiques of the dogmatic mentalism of the various insight therapies arose. The counter criticism was that behaviorism offered no clinically useful alternative. This situation was rectified in the 1950s with the development of the *behavior therapies*. When the cognitive revolution began in the 1950s, its critique of behaviorism also arrived without a practical clinical alternative. In the 1970s, *cognitive therapies* made their appearance.

Beginning in the 1950s, mostly due to advances in pharmacology, conventional medicine finally began to make headway in the treatment of mental disorders. In the 1990s, with the advent of brain-scanning techniques and the Human Genome Project, the *biomedical approach* has now positioned itself to make great leaps forward.

Also during the 1950s, two very different types of talk therapy arose, *group therapy* and *family therapy*. Both differed from the ordinary notion of medical treatment in that more than one person is treated at a time.

Reading Rules

When reading about approaches to psychotherapy, it is important to remember that even though there are commonalities in symptoms within a disorder, the effect on each patient is highly individualized. The way symptoms are expressed and the specific way they affect a person's well-being can vary greatly from person to person. No two people experience depression in quite the same way, nor will an individual experience two different episodes of depression in quite the same way. Therefore, a course of treatment needs to be tailored to the specific patient at the time of treatment. We will discuss the generalities of each psychotherapeutic approach, but understand that the experience of a specific patient will vary greatly.

> ## Critical Caution
>
> It is important to note that while the various therapeutic approaches are discussed separately in this chapter, they are not practiced in a vacuum. Modern psychotherapy is increasingly taking an integrative approach to treating patients.

CURRENT APPROACHES

Broadly speaking, therapies for psychological disorders can be classified into three groups: those derived from various psychological approaches, those derived from biology and medicine, and those wherein more than one person is treated at a time.

Individual Psychologically Based Therapies

Psychotherapy techniques are most easily categorized according to the school of thought from which they were derived. As mentioned previously, the psychodynamic approach led to insight therapies; the behavioral approach led to behavior therapies; and the cognitive approach led to cognitive therapies. Each type enjoyed predominance during the time period in which the school of thought was in favor. Insight therapies have continued to be popular for many conditions where newer treatments have yet to prove effective. Behavior therapies have found a niche with a number of conditions, including phobias, obsessive-compulsive disorder, developmental disabilities, and various problems due to brain damage (whether from accident, drugs, or disease). The behavioral approach has also had a substantial influence on the other psychologically based therapies. With the cognitive approach still popular, cognitive therapies continue to be explored.

INSIGHT THERAPIES

The *insight therapies* are based largely on the work of Freud. These therapies include *psychoanalysis, analytic psychotherapy, humanistic therapy*, and *psychodynamic therapy*. The term, *insight*, refers to the fact that the goal of therapy is to change the patient's views in order to change her behavior. These therapies (and some others) are also called *talk therapies*, because the principle therapeutic method is for the patient to talk with the therapist. Even at this level, we can see the importance of a distinction between theory and method for insight therapies.

Different Types of Insight Therapy

Strict Freudian psychoanalysis not only constrains the content of the talk in talk therapy, but also the form of the talk. The patient lies on a couch, with the therapist out of view. The patient then learns to free-associate, which will take months, talking without the normal constraints of staying on topic or interacting with the therapist. Two of Freud's contemporaries, Jung and Adler, each developed alternatives to psychoanalysis. Both rejected Freud's one-sided conversations as autocratic. Jung developed alternative modes of talking based on free association. It was Adler who began to use more ordinary conversation with the patient in therapy, focusing his therapeutic techniques more on what was said and less on how it was said. Almost all later talk therapies followed Adler's example. The differences between the many types of talk therapy that followed were theoretical, not methodological.

Insight generally refers to the understanding of one's own psychological processes. The theoretical differences between most insight therapies revolve around how those psychological processes are understood. The methodological differences revolve around how the therapist gets the patient to understand them. In the psychodynamic view, including the psychoanalytic, symptoms are believed to reflect unconscious conflicts and compromises among competing wishes, fears, and maladaptive ways of defending against unpleasant emotions. In the humanistic view, the causes of symptoms, while not necessarily more available to consciousness, are more related to one's view of one's self and one's relationship to others, rather than to one's motivations, emotions, and biological needs. Humanistic and other nonpsychodynamic therapies also tend to focus more on present problems and less on past conflicts.

Methods

Each type of therapy involves modifications of Freud's original technique, but we will focus on the commonalities. Some of these techniques are older, more formal, and less often used nowadays. However, the notion of insight and the process and purpose of talk therapy are best understood by learning about the more formal techniques.

The insight therapies use techniques based on both what has been said in a therapeutic session and what has not been said. *Free association* was the first therapeutic tool used to explore mental associations and unconscious processes involved in symptom formation. The therapist encourages the patient to say whatever is on her mind without self-censoring. However, the patient's *resistance* is a barrier to free association. The free association process draws the patient closer to the source of the problem, increasing her anxiety. The patient becomes motivated to avoid the anxiety and thus avoid talking about the source of the problem. This is an important source of information for the therapist. The therapist must be able to recognize and interpret the patient's unspoken resistance.

Another important source of unspoken information for the therapist is based on the relationship between the patient and therapist. Freud observed that patients tend to play out many of the same troublesome interpersonal scenarios with their therapist that they experience in their daily lives. This *transference* of thoughts, feelings, and conflicts from past and current relationships onto the therapist can provide a wealth of information.

Humanistic therapists developed different methods both for overcoming resistance and for uncovering interpersonal problems. Carl Rogers' *client-centered therapy* uses techniques based on unconditional positive regard. By responding positively, but truthfully, to whatever the client has to say, the therapist hopes to increase the patient's self-esteem to the point where he is more able to be honest about even the most difficult topics. Fritz Perls' *Gestalt therapy* uses conversational techniques, including the empty chair and two-chair techniques, to allow the patient to present interpersonal conflict without the need for transference. The patient pretends that the other person with whom she has problems is sitting in the empty chair and confronts him. The empty chair can be a real person or, in the two-chair version, an alternative course of action.

The insight therapies also share a common understanding of what constitutes therapeutic change. First, therapeutic change requires that the patient comes to understand the internal workings of his mind. This means gaining the capacity to make conscious, rational choices as an adult about wishes, fears, and defensive strategies that may have been forged in childhood. This is not, however, a purely cognitive act. It emphasizes "emotional insight," or the principle that intellectually knowing about one's problems is not the same as confronting intense feelings and fears. Second, the relationship between the patient and therapist is critical for therapeutic change to occur. A patient needs to feel comfortable with the therapist in order to speak openly about emotionally significant experiences. Finally, by using both spoken and unspoken information, the patient and therapist can collaborate to "solve the mystery" of the symptom by piecing together the connections between both types of information.

Fun Facts: The Effect of Insurance on Talk Therapy

Movement away from the harder-to-learn and slower techniques invented by Freud allowed all sorts of talk therapists to develop treatment regimens that took less time (months instead of years) to address many psychological problems. This was critical, in part because insurance companies, reluctant to pay for many sessions, established restrictions on how many psychological treatments they would reimburse. (These sorts of restrictions do not exist for the treatment of physical ailments.)

BEHAVIORAL THERAPIES

In the 1950s and early 1960s, a number of psychologists rejected insight therapies as a form of treatment. They argued that they were based on unsupported principles and techniques that had not been scientifically validated. They turned to behaviorism and learning theory and viewed psychological problems as maladaptive learned behavior patterns.

These therapists are not concerned with exploring underlying personality patterns or unconscious processes. The focus is on the patient's behavior in the present. Therefore, they take a rational problem-solving approach and focus on the symptoms instead of insight. In addition, because their techniques are based on work with nonhuman animals, there is much less need for talk, which is an important advantage in dealing with nonverbal patients, including children, the developmentally disabled, and those with severe emotional or neurological problems.

Classical Conditioning Techniques

Today, therapeutic techniques based on classical conditioning are used primarily to treat phobias. The assumption is that through classical conditioning, a person has learned to fear what should be a neutral stimulus. For example, a person who has been stung by a bee feels afraid to go outside because being outside (the *conditioned stimulus*, or *CS*) is associated with a terrifying experience (the *unconditioned stimulus*, or *UCS*). Normally, future encounters with the CS (going outside) without it being paired with the stimulus that elicited the fear response (the bee) will lead to the extinction of the response (fear). However, if the person avoids the CS (going outside), this short-circuits the adaptive learning process. The phobic person avoids the fear by not going outside, but this prevents extinction, so the fear remains. This vicious cycle of avoidance and fear produces the persistent impairment characteristic of phobias.

If a phobia can be classically conditioned, therapeutic techniques based on counter-conditioning should alleviate the phobia. *Systematic desensitization* is one such technique. The first step is to teach the patient various relaxation techniques, such as tensing and relaxing various muscle groups. The therapist then questions the patient about her fears and uses this information to construct a hierarchy of feared imagined stimuli that ranges from mild anxiety to intense panic.

For example, the patient might only experience mild anxiety in response to the thought of a bee in a flower bed across the street, but experience greater anxiety when thinking of a bee hovering outside her window. A panic attack might result from the thought of a beehive. The desensitization process begins by having the patient relax and imagine the least threatening stimuli. While the patient imagines this, relaxation techniques are used until the imaging stimuli no longer causes fear. This process continues for each imagined stimulus up the hierarchy until the patient

Something Extra: Exposure Techniques

The *exposure techniques*, such as flooding and graded exposure, are similar to systematic desensitization except that they utilize real stimuli instead of imagined stimuli. Flooding involves exposing the patient to the fear object immediately and is based on the idea that inescapable exposure eventually desensitizes the patient through extinction. Graded exposure is systematic desensitization using a real fear object.

is confronted with the fear object itself. The key in systematic desensitization is to gradually break the learned association between the fear object and the fear response until that response is extinguished.

Operant Conditioning Techniques

The operant conditioning techniques use rewards and punishments to modify unwanted behavior. You will often see operant techniques used in smoking cessation programs and behavioral modification programs. These therapies are based on the idea that increasing the reward for a wanted behavior will increase the likelihood that the desired behavior will occur. Conversely, increasing the punishment for an unwanted behavior will decrease the likelihood that the unwanted behavior will occur. For example, smoking addiction has both biochemical and psychological components. While most over-the-counter cessation programs address the biochemical addiction, the most successful programs also incorporate behavioral modification techniques to address the psychological component. These techniques commonly use both punishment and reward. For example, a rubber band around the wrist that is snapped each time a person wants a cigarette provides a slight punishment to discourage the craving. In addition, putting $5.00 in a jar each time you do not buy a pack of cigarettes acts as a reward for not making the purchase. Combined, the punishment decreases the likelihood of the smoking behavior, and the reward increases the likelihood of not replenishing the supply.

Observational Learning Techniques

The observational learning techniques are based on the idea that people learn by observing the behavior of others. These techniques are often used in social skills training, which focuses on helping people cope better with specific types of social or work situations. For example, many people become overly anxious when faced with the prospect of speaking in front of an audience. A therapist using observational learning techniques would first model basic relaxation techniques to relieve the anxiety. Next, the therapist would model specific behaviors useful in public speaking, such as

Handy Hints

We should not take the connections between specific types of learning theory and specific therapeutic techniques too seriously. Classical, operant, and observational learning are part of almost every human behavior. For instance, could you describe systematic desensitization or social skills training in terms of operant conditioning? Behavioral techniques share a common source in laboratory practices of conditioning, irrespective of what mix of learning theories one uses to describe them.

making eye contact, modulating the voice, or using hand gestures. The therapy serves a dual purpose. The first purpose is to teach specific skills relevant to the situation. The second purpose is to make the person more comfortable participating in the situation.

COGNITIVE THERAPIES

In many cases, it is not just our overt behaviors that need to be changed in order to better cope with the world. The thoughts that lead to those maladaptive behaviors may also require change. The cognitive therapies focus on changing these dysfunctional thoughts, or cognitions, that may underlie psychological disorders. It is important to note that the cognitive therapies are often combined with other therapeutic techniques, as with the cognitive-behavioral therapy discussed later in this section.

Cognitive Therapy

The cognitive therapeutic techniques discussed here target the things people spontaneously say to themselves and the assumptions they make when interpreting their experiences. The goal of the techniques is to challenge and change these assumptions in the service of improving a person's well-being. Rational-emotive therapy is based on the assumption that if the source of psychological distress is irrational thinking, then the path to eliminating symptoms is increased rationality. The goal of the therapy is to challenge the irrational thoughts and create new, valid rational thoughts in their place.

For example, a student who is habitually late with his assignments may irrationally blame everyone but himself for their lateness (for example, "The dog ate my homework," "My computer crashed and I didn't have a backup," "My roommate kept me up all night"). A rational-emotive therapist would first challenge the irrational interpretation of the situation ("Do you really have a dog?" "Does he really eat paper?"). Once challenged, the therapist would help the student to interpret the situation in a more rational manner ("Did you plan enough time to complete your assignment?"). Once a more rational interpretation is reached, steps can be taken to decrease the problematic behavior.

Another technique from cognitive therapy is to take an empirical approach to changing irrational thinking. Beck's form of cognitive therapy assumes that irrational thoughts stem from cognitive distortions of reality. The way we interpret situations is systematically biased. In order to eliminate the bias, we should treat our interpretations as hypotheses to be tested. Thus, the therapy is a process of collaborative empiricism, in which the therapist and the patient work together like scientists testing hypotheses. To continue the habitually late student example, the therapist would have the student keep a daily log of his work on an assignment, his feelings about that work, and the thoughts he has about the work. The student may report that when he sits down to write a paper, he feels anxious because he doesn't think he will get an A, so why bother trying? The therapist will then encourage the student to restructure his thinking with the fact that a grade of B or C is better than an F. By adding this piece of "evidence" into the student's thinking, the student is less likely to procrastinate on the next assignment.

Cognitive-behavioral Therapy

The cognitive-behavioral therapies are a hybrid of both cognitive and behavioral therapeutic techniques. The assumption behind these techniques is that cognitive therapy can help change irrational thoughts, and those changes lead to changes in behavior. Behavioral techniques can help change behaviors, and those changes lead to new ways of relating to new experiences. Relating to new experiences can help a patient develop more rational thinking. Thus, most modern therapists recognize that cognitions, behaviors, and feelings all influence not only each other, but also a person's overall well-being.

Big Background: Cognitive Therapies

Historically, cognitive therapies originated with the looser styles of insight therapies popular in the 1960s. Cognitive therapists shared the behaviorists' frustration with the theoretical focus of both psychodynamic and humanistic therapies. Like the behaviorists, they also sought a more practical and straightforward way of dealing with the patient's problem, but one that allowed them to take advantage of the power of language in reaching out to verbal subjects. Intriguingly, there is not a lot of evidence that cognitive-behavioral therapies are superior to strictly behavioral therapies, although they *are* generally more effective than strictly cognitive therapies. The bottom line may be that people, even behavioristic people, talk. Unstructured talk between the patient and either the therapist or others in her life may lead to better thinking.

Biomedical Therapies

Biomedical approaches have a long history, but the recent rise in pharmaceutical development has made this approach one of the more popular today.

The *biomedical therapies* assume that psychological disorders reflect pathology of the brain, particularly of the neurotransmitters that carry messages from one neuron to another. Biological treatments use medication to restore brain chemistry to as normal functioning as possible. If psychopharmacological treatments fail, then electroconvulsive or psychosurgery can be used.

PSYCHOPHARMACOLOGY

Psychopharmacology is the study of medications that act on the brain to affect mental processes, mood, and behavior. Some medications act at neurotransmitter sites to inhibit overactive neurotransmitters or receptors that cause the neuron to fire too frequently. Other medications increase the action of neurotransmitters that are underactive or in too short supply. In many cases, the precise mechanism and its relation to behavior are not yet well understood. Because the field of psychopharmacology is growing rapidly, we cannot list every class of medication here. We will, instead, discuss three major classes of psychotropic medication: *anti-psychotics, mood stabilizers*, and *anti-anxiety medications*.

Anti-psychotics are used to treat schizophrenia and other acute psychotic states. The hallucinations and delusions experienced by patients are believed to result from an excess of dopamine available in the brain. The anti-psychotic medication inhibits the activity at dopamine receptors, thus reducing the hallucinations and delusions of psychosis and schizophrenia. It is important to note that anti-psychotic medications are not a cure for schizophrenia. They only decrease some of the prevalent symptoms.

Mood stabilizers are used to treat depression, bipolar disorder, and some personality disorders. These medications are commonly referred to as antidepressants, but are often used to treat more than just depression. Most mood stabilizers target the neurotransmitters norepinephrine and/or serotonin, both of which are depleted in depression. By increasing the availability in the brain of one or both of these neurotransmitters, they can reduce depressive symptoms. There are four subclasses of mood stabilizers. The *tricyclic* antidepressants block the reuptake of serotonin and norepinephrine and cause the neurotransmitters to stay in the synapse longer. *Monoamine oxidase* (MAO) *inhibitors* prevent the chemical MAO from breaking down neurotransmitter substances, which allows more neurotransmitter to be

Critical Caution

Recent discoveries in the research laboratory about serotonin have given rise to much confusion about the mechanism by which some of these medications work. This fact emphasizes both that psychopharmacology is in its infancy and that these medications are discovered on an empirical basis of what works, rather than from a purely neurophysiological theory. For example, benzodiazepines were in wide use as anti-anxiety agents long before GABA was even discovered.

available for release into the synapse. *Selective serotonin reuptake inhibitors* (SSRIs) work much like tricyclic antidepressants, but target only serotonin. Finally, *lithium* is the treatment of choice for bipolar disorder. It is used to equalize mood by bringing the patient out of a manic or depressive phase and to prevent future mood swings.

Anti-anxiety medications, or benzodiazepines, are prescribed for short-term treatment of anxiety symptoms, such as panic attacks. These medications increase GABA activity, which is a neurotransmitter that inhibits activation throughout the nervous system. Reducing this activation reduces anxiety. It is important to note that most psychiatrists also prescribe antidepressants to patients who suffer from anxiety disorders. While this may seem counterintuitive, the neurotransmitters involved with depression also serve many other functions.

ELECTROCONVULSIVE THERAPY

Electroconvulsive therapy (ECT) is a last-resort treatment for severe depression, some forms of mania, and some obsessive-compulsive disorders. ECT, or "shock treatment," has a controversial history in the treatment of psychological disorders and has largely fallen out of favor in all but the most extreme or untreatable cases. Developed in the 1930s, ECT is generally conducted to induce a controlled brain seizure. It was assumed that because schizophrenia and epilepsy could not coexist, inducing an epileptic seizure via ECT could treat symptoms of schizophrenia. Today, we know that the fundamental assumption is inaccurate, and ECT is no longer considered a treatment for schizophrenia. The mechanism by which ECT is effective in treating depression is not known.

Despite a history of being used irresponsibly, ECT can be an effective treatment when other methods fail.

Big Background: ECT

In part because it is an older therapy, and in part because it does not involve medications (which would place it under the auspices of the Food and Drug Administration), ECT has received much less scrutiny than other biomedical therapies. Over the decades, public pressure due to revelations about misuse and overuse of ECT has led both to modifications of the technique and its use with fewer disorders. Lower voltages, the use of anesthesia and muscle relaxants, fewer treatments, longer waits between treatments, stricter stopping rules, and the use of unilateral (one side of the head only) rather than bilateral shock have all become more common. Despite its dramatic short-term effects on depression, depression often recurs a few months after treatment. Its effectiveness with other types of disorders is not well-established.

PSYCHOSURGERY

Psychosurgery is the chemical or surgical removal of a part of the brain. Originally, the most common form of psychosurgery was the lobotomy, which involved damaging or removing a part of the cerebral lobe. By the 1950s, these techniques were used to treat psychotic, violent, or difficult patients. Unfortunately, the side effects rarely outweighed the benefits. Patients often became apathetic, could not think abstractly, or lost self-control. The lobotomy is no longer used for psychological disorders. More recently, psychiatrists have been studying more limited forms of surgery to treat the severely debilitating cases of obsessive-compulsive disorders that don't respond to other treatments. Psychosurgery is also the only effective treatment for intractable seizures that do not respond to medication. These surgeries also can have serious side-effects.

Fun Facts: Ancient ECT and Psychosurgery

While it is easy to imagine that ECT and psychosurgery are relatively modern Western techniques, the concepts behind the procedures are really quite ancient. Hieroglyphics on the walls of Egyptian tombs depict the use of electrical eels to numb emotional states. Skeletal remains from thousands of years ago show holes bored into skulls, presumably to allow demons to escape from the heads of the mentally ill (although other skulls show similar holes that might have been effective treatments for relieving fluid pressure after head injuries). Even today, an indigenous tribe living off the coast of Panama uses a potion that induces convulsions to ease psychotic states. Thus, while the specific procedures of psychosurgery and ECT are relatively modern Western inventions, the *concepts* of psychosurgery and ECT are most definitely not.

Nonindividual Therapies

The *nonindividual therapies* also have a long history, but rose to prominence in the post–World War II era. Initially, the reason was economic. The increased demand for therapeutic services and the shortage of therapists (not to mention the cost of individual therapy) created the need to treat more than one patient at once. As these therapies developed, however, a deeper understanding of theoretical advantages to treating people in groups arose.

To this point, we have discussed therapy in terms of a one-on-one relationship between an individual therapist and an individual patient. One patient at a time is a nearly universal rule in medicine, after all. This *individual therapy* often works very well for patients. However, many of the therapeutic techniques discussed also lend themselves well to groups of patients being treated by an individual therapist. The insight, behavioral, and cognitive therapeutic techniques, either alone or in combination, are often administered to groups of patients simultaneously.

GROUP THERAPIES

Group therapies arose from two sources. Humanistic therapists were concerned about the lack of availability of therapy to less affluent patient populations. They began to lead groups of patients. Outside of psychology, self-help groups, most importantly, Alcoholics Anonymous, developed a leaderless model for groups.

Therapist-led Groups

There are a wide variety of group therapies, but they can be classified into two main types: *homogeneous groups* and *heterogeneous groups*.

Homogeneous groups consist of people who share a specific problem or psychological disorder. The demographics of the participants, the underlying causes of the problem, and the way the symptoms interfere with their lives may be varied, but the disorder is shared. There are several advantages of homogeneous group therapy over individual therapy. First, the shame, stigma, and isolation that often accompany a psychological disorder can be alleviated by interacting with others also living with the disorder. Second, a new patient to the group can see others in various stages of recovery or coping, giving the new patient hope that she will also make progress. Finally, patients within the group can share successful coping strategies or skills that aid the others in the group. (This is a good example of an important therapeutic advantage to group therapy over individual therapy.)

Heterogeneous groups, on the other hand, consist of people who do not share a particular disorder, but are interested in learning how to better interact with others. The goal of the therapy is to change maladaptive interaction patterns, but the

> ## Big Background: Relating to Others as Therapy
>
> The advantages to group therapy arose after the fact. Nonetheless, they fit in well with the humanistic notion that disruptions in interpersonal relations are keys to mental health. Bad interpersonal relationships can create problems, but good ones can help solve them.

patterns themselves vary widely within the group. In addition to the advantages just discussed, heterogeneous group therapy gives patients the opportunity to practice their new interaction strategies in a safe place.

Self-help Groups

Self-help groups, or support groups, are a form of homogeneous group therapy, but they are not directed by a mental health professional. In most cases, a therapist will recommend a self-help group to supplement individual therapy. The self-help groups are centered on a single disease, maladaptive behavior pattern, or traumatic experience, and are directed by those who have either recovered or developed successful coping strategies. Some of the better-known self-help groups include Alcoholics Anonymous, Weight Watchers, cancer survivor groups, and Gamblers Anonymous. Self-help groups afford most of the advantages of other homogeneous group therapies. The main disadvantage is that these groups tend to oversimplify the causes of the problem, leading patients to believe that the root of all their problems is their addiction, trauma, or disease. When coupled with individual therapy, this oversimplification can be remedied.

FAMILY THERAPIES

Family therapy provides treatment for either the entire family unit or various sub-sets of the family. For example, parents and children may attend therapy together,

> ## Key Point: Therapy for Teams
>
> Unlike group therapy, family therapy was developed from a theoretical perspective. In strong contrast to the medical model, systems theorists in the 1950s and 1960s were deeply committed to the notion that mental problems were often problems with a social unit, rather than an individual. This perspective has led to other sorts of interventions with other sorts of intact groups. For example, *retreats* for coworkers in an office or for teams in occupations such as police, medical, rescue, or the military can be used to "treat" the team and solve problems that the team has as a group.

Variant View: Integrating Therapeutic Approaches

One of the more interesting and important developments in the field of psychotherapy is the blending of different therapeutic approaches to treat an individual patient. Therapists recognize that different therapeutic techniques offer different advantages for the patient. For example, a therapist treating a patient with phobic disorder may use anti-anxiety medication (a biomedical therapy), systematic desensitization (a behavioral therapy), and rational-emotive therapy (a cognitive therapy). Thus, the therapist's theoretical orientation becomes less important in deciding a course of treatment.

or only couples or only siblings may attend therapy together. Nontraditional and blended families are also considered for family therapy. Essentially, any group of people that functions as a family is considered fair game for family therapy. There are also many approaches to family therapy. Many therapists will apply cognitive, behavioral, or insight therapies to the family dynamic. Here, we will discuss one approach that is relatively unique to family therapy.

Family systems therapy assumes that a family is a system of interdependent parts. The maladaptive situation lies in the structure of the system itself, rather than in the family member who is manifesting the symptom. The symptoms occur in the larger context of the family, and changing one aspect of the system will affect other aspects of the system. The goal of family therapy is to change maladaptive family interaction patterns. Thus, the process of interaction is almost as important as the content of what is said in a session.

Summary

We have seen how psychotherapy has advanced from prescientific notions, through a period of close linkage to various theories of the mind, to the current, more practical and empirical approaches. The recent advances in psychopharmacology, along with new genetic science that promises not only major leaps forward in diagnosis, but also individualized biomedical treatments customized to the patient's genetic profile, have led to enormous enthusiasm for strictly medical approaches to psychotherapy. We should remember, however, that the present enthusiasm for exclusively medical therapies is no greater than was the enthusiasm for strictly psychoanalytic therapies in the 1920s. Every sort of treatment has its limits. For newer treatments, those limits are yet to be discovered.

Perhaps the most important advance due to the influence of biomedical therapies is the increasing demand for the empirical validation of all therapies. Treatments, new and old, must prove themselves safe and effective. When more than one type of treatment is needed, the therapies can be selected from all of the proven therapies available, rather than merely from those fitting in with just one theory.

Quiz

1. The _____ therapies have the goal of changing the patients' views in order to change their behavior.

   (a) Biomedical

   (b) Insight

   (c) Behavioral

   (d) Cognitive

2. Which of the following is a source of information from the patient in insight therapies?

   (a) Free association

   (b) Resistance

   (c) Transference

   (d) All of the above

3. Classical conditioning therapeutic techniques are primarily used to treat _____.

   (a) Phobias

   (b) Schizophrenia

   (c) Relationship problems

   (d) Depression

4. Reward and punishment techniques used to help smokers stop smoking typically use _____ techniques.

   (a) Classical conditioning

   (b) Operant conditioning

   (c) Observational learning

   (d) Biomedical

5. _____ is based on the assumption that if the source of psychological distress is irrational thinking, then the path to eliminating symptoms is increased rationality.

   (a) Beck's cognitive therapy

   (b) Rogers' client-centered therapy

   (c) Rational-emotive therapy

   (d) Freud's psychoanalysis

6. Which of the following class of psychopharmaceuticals is not a mood stabilizer?

   (a) Monoamine oxidase (MAO) inhibitors

   (b) Selective serotonin reuptake inhibitors (SSRIs)

   (c) Lithium

   (d) Benzodiazepines

7. _____ group therapies consist of people who share a specific problem or psychological disorder.

   (a) Homogeneous

   (b) Heterogeneous

   (c) Family

   (d) Individual

8. Alcoholics Anonymous, Weight Watchers, and cancer survivor groups are examples of _____ groups.

   (a) Heterogeneous

   (b) Family therapy

   (c) Self-help

   (d) Therapist-led

9. _____ therapy assumes that the maladaptive situation lies in the structure of the family, rather than in the family member who is manifesting the symptom.

   (a) Self-help

   (b) Cognitive-behavioral

   (c) Psychoanalysis

   (d) Family systems

10. _____ is a last-resort treatment for severe depression, some forms of mania, and some obsessive-compulsive disorders.

   (a) Electroconvulsive therapy (ECT)

   (b) Psychopharmacology

   (c) Psychosurgery

   (d) Group therapy

PART ONE TEST

1. If we attribute the cause of a behavior as being _____ and _____, we are likely to say that the behavior is due to effort.

   (a) Internal, unstable

   (b) Internal, stable

   (c) External, unstable

   (d) External, stable

2. The _____ is the general tendency to prefer internal attributions over external attributions when explaining others' behavior.

   (a) Self-serving bias

   (b) Actor-observer bias

   (c) Fundamental attribution error

   (d) None of the above

3. Which of the following is a source factor in the study of attitude and attitude change?

   (a) Credibility

   (b) Attractiveness

   (c) Power

   (d) Sidedness

4. Which of the following is a compliance strategy based on the reciprocity norm?

   (a) Luncheon technique

   (b) Ingratiation

(c) Door-in-the-face technique

(d) Foot-in-the-door technique

5. _____ temperament types typically react negatively to new situations, usually by crying or fussing.

(a) Approach

(b) Avoidant

(c) Withdrawal

(d) Easy

6. Information-processing theorists view cognitive development as a series of _____ different stages.

(a) Quantitative

(b) Qualitative

(c) Both a and b

(d) None of the above

7. As children mature, their schemas change in which of the following ways:

(a) They become more articulated.

(b) They become more differentiated.

(c) Both a and b.

(d) None of the above.

8. _____ involves incorporating new information into an existing schema.

(a) Equilibration

(b) Assimilation

(c) Accommodation

(d) Articulation

9. Which of the following are characteristics of a "good" test?

(a) Standardized

(b) Valid

(c) Reliable

(d) All of the above

10. _____ reliability occurs when the identical test is administered on two occasions.

    (a) Split-half

    (b) Test-retest

    (c) Criterion

    (d) Alternate-form

11. In the equation IQ = (MA/CA) × 100, CA refers to _____.

    (a) Chronological aggregate

    (b) Chronological acuity

    (c) Chronological age

    (d) All of the above

12. _____ is the pattern of characteristic thoughts, feelings, and behaviors that persist across time and situations.

    (a) Emotion

    (b) Personality

    (c) Genetics

    (d) Cognition

13. According to Freud, _____ is/are tension systems created by the organs of the body.

    (a) Id

    (b) Ego

    (c) Drives

    (d) Superego

14. The id is driven by the _____ principle.

    (a) Pleasure

    (b) Reality

    (c) Morality

    (d) All of the above

15. The _____ is the storehouse of an individual's values, including moral attitudes learned from society.

    (a) Id

    (b) Ego

    (c) Superego

    (d) Personality

16. According to the Health Belief Model, what factor(s) determine whether or not an individual will alter his or her health-related behavior?

    (a) Perceived susceptibility

    (b) Perceived severity

    (c) Cues to action

    (d) All of the above

17. The Planned Behavior theory adds which factor to the Reasoned Action theory of health behavior?

    (a) Self-efficacy

    (b) Genetics

    (c) Cues to action

    (d) Perceived severity

18. _____ is a type of stress that involves demands or expectations to behave or perform in a specific manner.

    (a) Conflict

    (b) Frustration

    (c) Change

    (d) Pressure

19. _____ schizophrenia is characterized by delusions and hallucinations.

    (a) Paranoid

    (b) Catatonic

    (c) Disorganized

    (d) Undifferentiated

20. _____ are persistent unwanted thoughts or ideas; _____ are repetitive behaviors.

    (a) Compulsions; obsessions

    (b) Obsessions; compulsions

(c) Mania; depression

(d) Depression; mania

21. Which of the following is not a diagnosed personality disorder?

(a) Anti-social

(b) Depression

(c) Schizotypal

(d) Avoidant

22. Which of the following is not a treatment for phobias?

(a) Flooding

(b) Systematic desensitization

(c) Exposure

(d) None of the above

23. _____ group therapies consist of people who do not share a specific problem or psychological disorder.

(a) Homogeneous

(b) Heterogeneous

(c) Family

(d) Individual

24. _____ are used to treat schizophrenia.

(a) Monoamine oxidase (MAO) inhibitors

(b) Selective serotonin reuptake inhibitors (SSRIs)

(c) Lithium

(d) Anti-psychotics

25. Talk therapy is another term for the _____ therapies.

(a) Behavioral

(b) Biological

(c) Insight

(d) Cognitive

PART TWO

The World of the Mind

A great deal of what we do and why we do it is hidden from public view. Attempting to understand the private side of people is critical to understanding the public side.

Psychology literally means the study of the mind. While our understanding of just what the mind is has changed due to the influence of the scientific method in psychology, a number of areas of psychology are still devoted to understanding those aspects of human behavior that are so hard to observe. With the advent of new brain-scanning technologies, our understanding is growing in new ways. Part Two discusses what psychology has discovered about those things traditionally associated with the concept of the mind.

Chapter 9, "Emotion," goes into what psychologists have had to say about our feelings. Chapter 10, "Motivation," looks into the role of wants and needs in causing behavior. Chapter 11, "Learning," examines how experience changes us and makes us more effective in the world. Chapter 12, "Psycholinguistics: The Psychology of Language," talks about talking… and listening. Chapter 13, "Memory," investigates how we are able to access our own past experience. Chapter 14, "Cognitive Psychology: The Study of Thinking," covers what some people take to be the main purpose of the mind, thinking.

CHAPTER 9

Emotion

This chapter covers psychologists' efforts to understand emotions from a scientific perspective. We will discuss how psychologists attempt to categorize the different emotions, theories of how emotions come about and what purpose they might serve, and how emotions are communicated from one person to another.

Tips on Terms: Defining Emotion

Psychologists have no single, agreed upon definition of *emotion*. Different introductory psychology texts have different, often contradictory, definitions for *emotion*. There is nothing wrong, in principle, with conducting scientific research on something without first being able to define it. On the other hand, the fact that your introductory textbook gives one and only one definition of emotion does not mean that it is the definition used by all psychologists who study emotion.

Overview

Emotions are things we are all aware of through experience. We also have a language we use to talk about them. This makes emotions one of those topics in psychology that everyone already feels somewhat expert about. As usual, psychologists try to re-examine, from a scientific perspective, those things we already think we understand. More than most other areas of psychology, the study of emotion falls under Bode's (1922) rubric of talk about things everybody knows in terms no one understands (see Glossary). It may help for a moment to step back and think about what emotion is before we look at how psychologists study it.

Emotions are parts of our internal lives but they are not things that we do, and thus they do not always have immediate connections to events in the outside world. These sorts of happenings seem to have important consequences and a distinctive character. However, unlike thoughts, emotions seem more like things that happen to us and less like things we do.

Emotions seem to come in different flavors, and psychologists try to classify the different emotions. Emotions also seem to affect and be affected by our actions and our thoughts, and psychologists theorize about this as well. We have the ability to detect not only our own emotions, but also the emotions of others, at least to a degree. Finally, as with everything else psychological, we assume that the underpinnings of emotion are involved with bodily events and psychologists are concerned with determining which bodily events are involved.

HISTORY

Even before the scientific era, emotions were discussed and studied. In the West, there was not much concern with classifying the emotions. Plato and Aristotle, for instance, were mostly concerned with the functions of the emotions, which they thought involved promoting ethical behavior. Physicians such as Hippocrates were concerned with the processes that caused emotions. In the East, however, there were early attempts both in India and China to classify the emotions.

Key Point

Psychological studies of emotion tend to focus on (1) classifying different types of emotions, (2) understanding the processes that make emotions happen, (3) understanding the function or purpose of emotion, and (4) understanding how we communicate our emotions to one another.

The modern study of emotion began with Darwin and continued with Freud; both were concerned primarily with the function of emotions. Unlike the ancient Greeks, Darwin thought that the function of emotions was survival, not ethics. Emotions helped us avoid bad things and approach good things. With the advent of behaviorism, a problem arose. Approach and avoidance could now be studied without worrying about what emotion was being experienced. Some behaviorists referred to emotions as *epiphenomenal*, meaning that they were evolutionary side-effects of the actual, effective processes that governed approach and avoidance and other observable behaviors. As a result, contemporary research into emotion is one of the few areas of psychology where most research tends to neglect the issues of function.

CURRENT APPROACHES

Neglecting function, contemporary psychological research tends to focus on classification and process, with some more recent research on communication. Most classification research is based on the notion that there are a few basic emotions, with the remaining emotions being variations. Early process research began by focusing on the relationship between feeling an emotion and identifying it. These researchers tie their notions of process closely to the question of classification. With the more recent advent of sophisticated techniques for measuring physiologi cal activity, including brain activity, process research has begun to focus more and more on the physiological processes underlying emotions. More advanced measurement techniques have also created a boom in communication research.

Types of Emotions

The first thing to understand about how psychologists attempt to classify emotions is that our ordinary names for emotions are already a classification system. This system (or systems, if you think about different names in different languages, such as English or Hindi or Chinese) was obviously not designed for scientific purposes. Instead, it evolved as part of our need to describe our different emotions. As you might expect, different languages have different structures, and even different gaps in their system. There are languages that have names for emotions that have no names in other languages. There are languages that do not distinguish between emotions that have multiple names in other languages.

Despite these problems with preexisting natural classification systems, in psychology, almost all classification research on emotions begins with the names of the emotions. In the English-speaking world, this type of research usually begins with the English names. In order to understand how psychologists classify emotions, we need to take a look at some of the peculiarities of English names for emotions.

The first thing to understand is a distinction in English that is probably not mentioned in your introductory psychology text, the distinction between *sentiments* and *moods.* Sentiments (sometimes called *emotional attitudes* by psychologists) are emotions that are directed to or created by things in our environment, such as situations, events, objects, and people. Moods are more pervasive emotional states that are not obviously focused on any one thing.

Research into classifying emotions usually involves trying to build a hierarchical system. Certain emotions are labeled "basic," and all the other emotions are fit into those basic categories as variations on those few basic emotions. Most classification systems agree on four basic emotions, *joy*, *sadness*, *anger*, and *fear*. (The most common fifth emotion is *disgust*.)

As you can see, we are already having a problem with language. The first two of these four basic emotions are the names of moods, and the last two are the names of sentiments. With a little bit of work, we could find more or less equivalent names so that we had four sentiments or four moods, but psychologists do not usually do this. As you will see, we will get a better understanding of emotions if we do not do this either.

It is a bit easier to begin by thinking about sentiments rather than moods. First, we divide up the things that we get emotional about into those that give us pleasure and those that give us pain. (These are functional categories used by Plato, as well as by Darwin and Freud.) In the case of pleasurable things, the situation is simple: The presence of something pleasant makes us happy; its absence makes us sad. In sentimental terms, we love or like pleasant things, and we desire, or yearn for, or long for absent pleasant things. (Some systems classify joy and happiness separately from love, but this is most likely because they ignore the difference between sentiments and moods.) In biological or behavioral terms, we tend to *approach* pleasant things.

In the case of painful things, the situation is a bit more complex: In the absence of something unpleasant, we may not feel any emotion at all. In the presence of something painful or unpleasant, we have at least two options: We can try to get away from the unpleasant thing, or we can try to change it. Broadly speaking, in

Tips on Terms: Moods

Your psychology text may define *mood* more narrowly, as an emotional state that lasts only for a brief time. For instance, being in love would not be counted as a mood if it lasted a long time. Here, we will use the term *mood* in the broader sense.

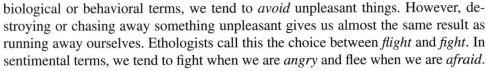

biological or behavioral terms, we tend to *avoid* unpleasant things. However, destroying or chasing away something unpleasant gives us almost the same result as running away ourselves. Ethologists call this the choice between *flight* and *fight*. In sentimental terms, we tend to fight when we are *angry* and flee when we are *afraid*.

Note that, with the exception of words such as *love*, English does not give us a lot of language to talk about sentiments regarding pleasant things or about moods regarding unpleasant things. *Love* and *like* are sentiments paired with the moods, *joy* and *happiness*. But the sentimental pairs for *sorrow* and *sadness*, for instance *yearning* and *longing*, are less common words and rarely appear as basic categories. (The situation is made even more complex by the fact that, if we think of *hope* as an emotion, then it is at least as good a mood to match with *yearning* as *sadness* is. But *hope* and *sadness* are clearly different moods.) The word for the mood corresponding to *fear* is *anxiety*. This is also an uncommon word and has been taken over by psychologists in a very technical sense. The only mood word corresponding to *anger* would be *rage*, and *rage* is also used to name a sentiment, meaning extreme anger.

English names have probably evolved asymmetrically because our behavioral options are not symmetrical. We love pleasant things and hate unpleasant things. But the presence of pleasant things does not always spur us to action. Instead, it may just make us feel good. And the only action spurred by the absence of pleasant things is to do something to make them pleasant. In neither case do pleasant things spur us to *specific* action. Thus, most of our emotion words about pleasant things are general terms for describing our moods, indicating how we feel. On the other hand, unpleasant things spur us to specific actions regarding that thing. So most of our emotion words about unpleasant things are specific terms about our sentiments toward those things, indicating what we might do.

In this context, we can think about *disgust*. Disgust seems to fall between anger and fear. Disgust is directed toward unpleasant things we either need not or cannot destroy or escape. Instead, we tend to arrange to get them out of our lives as quickly as possible. Certain unpleasant things, like dirt and grime, are inevitable parts of life. We cannot destroy them entirely, so anger is useless. We cannot escape them completely, so fear is also useless. The purpose of disgust is to ensure that we accept their arrival with some good grace, and then either destroy them, flee from them, or throw them away, as quickly as possible.

Extra Exercise

Do you think that feeling hopeful is an emotion? If you were building a classification system for emotions, would you make *hope* one of the basic emotions? If not, why not?

As you can see, a classification scheme that takes the difference between sentiments and moods seriously leads us to a consideration of function. It may be that the reason that contemporary research on emotion neglects this difference is because behaviorism challenged the functional analysis of emotion so effectively. In order to avoid problems with function, classification schemes neglect the difference between sentiment and mood.

Theories

Emotion is an area of psychology where theories abound. Psychologists were studying emotion long before theorizing in psychology became unpopular. On the other hand, these early theories may look a bit odd. After all, we have just been talking about different types of emotions. Shouldn't theories of emotion explain what makes sadness different from happiness, or what makes fear different from sadness? Instead, as we can see from Figure 9-1, theories of emotion talk about the causes that all emotions have in common.

TRADITIONAL THEORIES OF EMOTION

Traditional theories of emotion actually address both issues. By identifying the different elements that make up all emotions, these theoreticians sought to pinpoint the

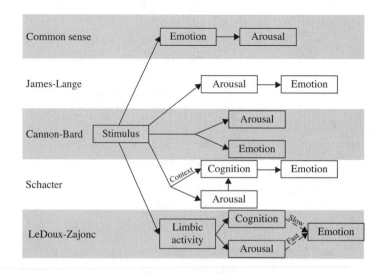

Figure 9-1   Theories of emotion

Big Background: Defining Emotion

Another confusing thing about theories of emotion is that, if your introductory textbook gives a definition of emotion, you may notice that not all of the traditional theories make sense with that definition. For example, some definitions may make arousal a cause or effect of emotion and others may make arousal a part of emotion. This is because the theorists, like the textbook authors, didn't all agree about exactly what emotion is. It is also because of an older, prescientific idea (still common in psychology) that theories have to be about causes and effects. As we will see, this focus on causes can create real confusion, because emotions are *both* causes and effects of other important psychological phenomena.

sources of the things that make different emotions different. Before we had even our present, very limited understanding of the brain activities that underlie emotion, psychologists examined the physiological events surrounding emotion. They looked to factors such as increased heart rate, sweating, and muscle tension as "arousal." Because arousal seems to happen at the same time as almost all emotions almost all of the time, the question seems to be what role does arousal play in causing emotion? However, because arousal is common to all the emotions, something else seems to be needed to explain what makes the different types of emotion different. The various traditional theories of emotion give different answers to these questions.

Let us take a quick look at the differences between the various traditional theories: Because emotions seem to be something that happens to us more than something we do, perhaps it is our arousal that is causing us to have emotions. In Figure 9-1, this is listed as the common sense view. Intriguingly, it has no advocates among the psychological theorists. The history of psychological theories of emotion is a movement away from this common sense view.

As we have mentioned a number of times, throughout the twentieth century, many psychological theories were considered to be incompatible alternative views, only to be shown later to be complementary, each describing different parts of what is going on. The James-Lang and Cannon-Baird theories were originally viewed as competing theories. However, we can see from the diagram that many of the differences among theories of emotion have to do with expanding our view of additional ways emotions can happen and the details of how they happen.

Recent neurophysiological research suggests that processes similar to both those envisioned in the James-Lang and the Cannon-Baird theories occur in the brain. The supposed contrast between these two was that the James-Lang theory required that different emotions required different types of arousal, whereas the Cannon-Baird theory assumed that there was only one kind of arousal that led to the various emotions. But even this contrast is not necessarily the case. The Cannon-Baird process would still work with multiple kinds of arousal.

Schacter agreed with Cannon-Baird that there was only one kind of arousal, in part because he believed that it was stimulus information separate from the primary stimulus (what he called *context*) that provided the information as to what emotion we experienced. More recent cognitive theorists accept the neurophysiological evidence that there are different types of arousal. However, this does not mean that there are enough different types of arousal to account for all of the different types of emotion.

A more important feature of cognitive theories, including Schacter's, is that what they call cognitive appraisal is really an elaboration of the arrow seen in the diagram of the James-Lang theory that goes from arousal to emotion. The James-Lang theory argues that arousal is a cause of emotion. Part of Schacter's view is to explain *how* arousal causes emotion, via cognitive processing.

LeDoux and Zajonc identify the notion of cognitive processing with cortical activity, which takes too much time to account for our quickest emotional reactions. (See Chapter 17 for more details on various brain structures.) In fact, some aspects of emotional reaction actually precede autonomic arousal. The slower process (represented by the top path in the LeDoux-Zajonc part of the diagram) corresponds to Schacter's expanded view of the James-Lang theory. The faster process resembles both the Cannon-Baird process and the common sense notions, because LeDoux and Zajonc see both arousal and the brain activity (in the hypothalamus) as being part of emotion, rather than a cause of it.

Thus, we can see that the current views capture both of the original theories, but with some confusion about what it is that we should define as emotion and cognition. Lazarus has criticized the LeDoux-Zajonc view because the faster process must also involve some sort of appraisal of the stimulus information, even if that appraisal really occurs only in the amygdala. Why should we call the appraisal "cognitive" only if it occurs more slowly and in the cortex?

Finally, note that, historically, the theories have progressed from simple causal relations, through descriptions of possible causal processes, to the identification of brain activities. Despite all this progress, none of the current theories addresses the obvious: The various things called emotions by psychologists involve complex sequences of interacting brain activity at various levels. Once the first few seconds have passed, our initial reaction is complicated by arousal, emotion, cognition, environmental cues, bodily states, and so on, which all affect one another. It is most likely rare that there is a simple sequence of causes and effects corresponding to the arrows in the diagram.

For example, most of us, at one time or another, have probably been in situations where we wanted to control our emotions. Brain research has confirmed that people are able to do this. In this case, cognition overrides emotion, a causal relation that does not appear as an arrow in any of the theories in Figure 9-1. Furthermore,

Variant View: One-way Arrows

An important discovery based on brain research is that all of the parts of the brain involved in emotions send signals back and forth from one to the other. There are no one-way paths corresponding to the arrows shown in Figure 9-1. Why don't psychological theories use two-way arrows? There are two basic reasons.

First, twentieth century psychology was based on the notion of the stimulus as cause and the response as effect. As John Dewey pointed out as early as 1896, this led psychologists to neglect all of the obvious cases where psychological phenomena proceeded in closed loops, rather than in one-way reflex arcs. Second, there is a prescientific idea that the only proper role for a theory is to identify the causes of the phenomena in question. This focus on causes can create real confusion, because emotions can be both causes and effects of other psychological phenomena. Note that even the evolutionary theories, which talk about the role of the *effects* of emotions, are constructed in terms of causes such as selection pressures.

---

research demonstrates that people in different cultures control different emotions differently depending on the social rules in their societies. Some emotions are overridden in some contexts, other emotions in other contexts, and these relations differ from culture to culture.

Perhaps we do examine various subtleties of our arousal to determine what emotion we are experiencing. But this does not eliminate the possibility that cognitive activity cannot alter our arousal. When we encounter a threatening situation, we may flee or fight. Perhaps the intensity or quality of that initial arousal tends us to one or the other. But that tendency could be overridden moments later as cognitive activity takes into account environmental cues. Differences in brain activity may be differences in types of initial emotional response, or the brain activity may be the first steps in cognitive appraisal. The addition of brain studies to the study of emotion gives us more data, but not necessarily any more definitive answers.

EVOLUTIONARY THEORIES OF EMOTION

The evolutionary perspective is the basis for many different theories in the study of emotion. The basic question posed by the evolutionary perspective, which is mostly neglected by other researchers in emotion, is what function does emotion serve? Why are there emotions at all? The evolutionist answer is that emotions evolved for practical purposes of survival. Fear is related to flight. Anger is related to attack. Love and liking are involved with social relations.

Reading Rules

If emotion has an evolutionary purpose, then emotion must have a role in producing behavior that can improve the chances of our survival or the survival of our children. Emotion might be the cause of the behavior, or a side-effect of the real causes, or a system that ensures that our future behavior can be communicated to others, or it might be a cause of *motivation*, which we will discuss in the next chapter.

In terms of cross-cultural studies, the evolutionary perspective leads theorists to predict similarities across cultures because emotions have proved useful for a long time. In terms of identifying the basic emotions, the evolutionary perspective suggests that the basic emotions are those with survival value. For example, jealousy may have value in ensuring that one's genetic material is carried on into the next generation. The evolutionary perspective does not have as much to say when it comes to the processes that are the focus of the other theories discussed earlier, although we can make sense of some of the data from an evolutionary point of view. For instance, it makes sense that circuits for fear and anger would evolve so that those reactions could happen quickly enough to deal with dangerous situations.

Expression

Another interesting aspect of emotion is that we are pretty good at telling what emotions other people are feeling. People are said to *express* their emotions, which means they behave differently when they feel differently, and we, as observers, are often able to perceive and identify what emotions they are feeling. In fact, given that arousal prepares us for fleeing, flying, feeding, or sexual behavior (a very old joke), it could be argued that the role of emotion is to signal that preparedness. Emotion first informs us ourselves that we are amped up for a particular sort of behavior. Our emotional expressions then serve to inform others.

FACIAL EXPRESSION

The most obvious aspect of emotional expression is the change in facial expression that accompanies emotion. Not only are facial expressions located so as to be easily seen by others, they also seem to be distinctly different for different emotions. This leads to another consideration. The James-Lange theory, unlike the Cannon-Baird

theory, requires that the bodily changes that precede emotion be different for different emotions. James and Lange thought in terms of physiological arousal, but there is no reason that the bodily changes that allow us to distinguish between our various emotions might not be facial changes. Ekman and others have investigated the role that facial expressions may have, not only in expressing different emotions, but also in causing them.

NONVERBAL COMMUNICATION

In the 1970s, popular books taught people how to read emotional expressions found in parts of the body other than the face. This sort of expression was called "body language." The popularity of these books points out two things: First, emotions are expressed using parts of the body other than the face. Second, people tend to be less expert at reading nonfacial expressions than facial ones.

Because nonfacial expressions are less obvious, people may not be as careful or as effective when trying to control them. When we seek to hide our intentions, we may alter our facial expression, but leave our body language alone. Also, we appear to be better trained at spotting threatening body language than other kinds. This may have survival value, since people are more likely to fake a smile when their intentions are bad than they are to fake a frown when their intentions are good.

Unlike facial expressions, nonfacial expressions have not been investigated closely to determine whether or not they are due to genetic factors. It appears, however, that not only do people from different cultures use different body language, but there are individual differences as well.

CULTURE

Are we able to determine what other people are feeling because there are genetically determined bodily reactions to different emotional states, or because we share culturally determined emotional signals with those other people? Questions like this have spurred cross-cultural studies. The general upshot is that many basic reactions, such as smiling and frowning, seem to be shared not only across cultures, but even with people, such as the blind, who lack the ordinary opportunities to learn the usual reactions from others. Those deaf from birth have difficulties with certain aspects of speech pronunciation. Those blind from birth have facial expressions almost indistinguishable from the normally sighted.

There are, of course, cultural differences in emotional expression. It is easier to determine the emotional state of someone who shares our cultural heritage. One intriguing difference between cultures is the matter of emotional control. Recent experiments using brain scanning have shown that, when we are asked to control our emotions, we do not just suppress our emotional expression, but actually lessen the degree that we are feeling what we are feeling. When and under what circumstances we exert this sort of control differs from culture to culture.

Summary

The psychological study of emotion focuses on four areas of understanding. First, psychologists classify the different types of emotions. Using ordinary language as a basis, they build hierarchical systems with *joy*, *sadness*, *anger*, and *fear* as the basic emotions. Second, psychologists try to understand the processes by which emotions occur. The various theories examine the relationships between the stimulus, arousal, cognition, brain activity, and emotion. Third, psychologists try to understand the purpose of emotions. The evolutionary perspective argues that emotions evolved for the practical purpose of survival. Finally, psychologists try to understand how we communicate our emotional state to others and how we interpret the emotional state of others. Research focuses on how facial language and body language aid this communication.

Quiz

1. Psychological studies of emotion focus on which of the following?
   (a) Classifying different types of emotions
   (b) Understanding the processes that make emotions happen
   (c) Understanding the function or purpose of emotion
   (d) Understanding how we communicate our emotions
   (e) All of the above

2. _____ are emotions that are directed to or created by things in the environment; _____ are pervasive emotional states that are not obviously focused on any one thing.
   (a) Sentiments; moods
   (b) Moods; sentiments
   (c) Sentiments; emotions
   (d) Emotions; moods

3. Which theory of emotion requires that different emotions require different types of arousal?

   (a) Cannon-Baird

   (b) Schacter

   (c) LeDoux-Zajonc

   (d) James-Lang

4. Which theory of emotion equates cognitive processing with cortical activity?

   (a) James-Lang

   (b) LeDoux-Zajonc

   (c) Schacter

   (d) Cannon-Baird

5. Evolutionary theories of emotion tend to focus on the _____ of emotions.

   (a) Process

   (b) Structure

   (c) Function

   (d) Classification

6. When trying to identify someone else's emotional state, we are best able to "read" their _____?

   (a) Facial expressions

   (b) Mind

   (c) Body language

   (d) None of the above

7. The way psychologists first attempt to classify emotions is to use _____ because it is already a classification system.

   (a) Hierarchical structures

   (b) Sentiments

   (c) Moods

   (d) Ordinary language

8. Which of the following is not one of the four basic emotions?

   (a) Joy

   (b) Spite

(c) Sadness

(d) Fear

9. Most of our emotion words about pleasant things are general terms for describing _____; most of our emotion words about unpleasant things are terms for describing _____.

(a) Moods; sentiments

(b) Approach; avoidance

(c) Sentiments; moods

(d) Avoidance; approach

10. Cross-cultural studies of emotion have found that the basic emotional reactions of _____ are shared across cultures.

(a) Emotional control

(b) Heart rate

(c) Smiling

(d) None of the above

CHAPTER 10

Motivation

Motivation is the topic in psychology where we ask *why*: Why do people do the various things they do? The notion of motivation is that there are certain activities with special biological importance, and we engage in other activities in order to improve the odds of either accessing or avoiding opportunities to do these special things. Opportunities for food and sex are important to our survival and the survival of the species, so we do things in order to obtain them. Threats to our safety and the safety of our families must be avoided or combated. In addition, there are subtler situations of importance, such as success on the job or time spent with others, that motivate us as well.

Overview

The notion of motivation has a role outside of scientific study. In our everyday lives, we speak of eating because we are hungry, drinking because we are thirsty, running away because we are afraid, kissing someone because we love him or her, and so on. There are two basic ways in which psychologists have tried to make

Topical Talk: Question for Discussion

What is motivation? Motivations are often associated with feelings. When we haven't eaten in a while, we *feel* hungry. Do we then eat because we are hungry, or because we *feel* hungry? In other words, are the feelings necessary to motivate us? After all, hunger is more than just feeling hungry. There are physiological changes. The stomach is empty. Blood sugar gets low, etc. After you have considered the case of hunger, consider the following: Do we go to the party because we are lonely, or because we *feel* lonely?

the concept of motivation into a scientific one. The first, more common approach, is to assume that motivations like hunger, thirst, fear, love, and hate are actually causes of behavior, and to attempt to figure out the details in the laboratory. The second, more skeptical, approach is to ask if the things we ordinarily think of as motivations are really the causes of behavior and to study the relation between motivations and behavior in the laboratory. This issue, and others like it, has led to very different approaches to the study of motivation.

HISTORY

Many areas of psychology are legacies from prescientific times. Emotion, cognition, and motivation were all identified as being part of psychology by the time of Plato. These fields of study have faced the challenge of validating the prescientific notions or of developing new ones to handle the same questions.

Early on in the history of psychology, motivation was a key concept in explaining behavior. First Freud and the psychodynamic theorists and then the behaviorists all took the view that explaining behavior meant finding out why people did things. The psychodynamic view held that there was a core of energy, the libido, which could only be dissipated by the appropriate biologically important actions. Some behaviorists shared this view, differing mainly from the Freudians in terms of how to investigate these questions. Skinner then proposed that, so long as the biology of motivation was not well understood, scientific inquiry into why people do what they do should focus on the behavioral patterns of motivated individuals confronted with specific opportunities to obtain things that would satisfy their needs. This was the concept of reinforcement.

As the psychodynamic and behavioral views lost influence in psychology, the focus shifted away from motivation. The question became: How do people do what they do? The study of motivation itself, already divorced from the study of behavior due to the influence of Skinner, became even more isolated from other aspects of psychology. While the study of motivation continues, its connection to the study of personality, learning, and other areas of psychology has lessened.

CURRENT APPROACHES

Historically, the different approaches to motivation often developed in conflict with one another. However, as we will see, these different approaches often focus on different aspects of motivation and are not as much in conflict as they might initially appear.

Psychodynamic Approach

The psychodynamic approach to motivation began with Freud and is one of the earliest psychological perspectives to theorize about human motivation. Freud proposed that, like animals, humans are motivated by two primal instincts, or basic drives: sexual instincts and self-preservation instincts. The sexual instincts motivate behaviors related to organ-pleasure, which develop into behaviors related to reproduction. The self-preservation instincts, or ego instincts, motivate behaviors related to our survival. These include obvious behaviors relating to hunger and thirst, but also include the ego's motivation to mediate conflicts between the id's desire for sexual satisfaction and the superego's desire for socially acceptable behavior. Thus, the sexual instinct and the ego instinct are in conflict.

Largely due to the influence of World War I, Freud later reorganized his categories of primal instincts to distinguish between sexual drives and aggression. He reasoned that because war kept "intruding" on "civilized societies," and strong proscriptions against killing have long been a part of our cultural, philosophical, and theological thinking, then there must be some instinctual drive humans were trying to prevent with these proscriptions. In other words, if aggression were not a part of human nature at an instinctual level, then why do we have to try so hard to prevent it?

The hallmark of the psychodynamic approach to motivation is the notion that the basic drives can exert their influence unconsciously. We are largely unaware when our instincts are pushing us in one behavioral direction or another. However, this does not mean that drives are always unconscious. Modern psychodynamic theorists argue that there are both conscious (explicit) and unconscious (implicit) motivations. For example, our sexual instinct may implicitly steer us toward a certain type of potential partner, such as tall brunettes or musicians. We can also deliberately seek out that type of partner in order to fulfill our sexual urges. The difference is only in whether or not we are aware of the motivation or drive underlying the behavior. Unconscious motivations are of more interest to psychodynamic theorists because the conflict between drives is believed to be largely unconscious.

Behavioral Approach

You will rarely hear a behavioral psychologist talk about motivation as an internal state. As discussed in Chapter 11, "Learning," the behaviorists do not posit internal

Variant View: Establishing Operations

Of course, strict behaviorists do not speak of hunger being the motivating factor, because hunger is internal and thus hard to observe. Behaviorists speak of prior external events, such as being deprived of food, as being the prior environmental contingencies, called *establishing operations*, that make food reinforcing. The value of the present contingency, the availability of food, is altered by the occurrence of the prior contingency. This is how behaviorists speak about motivation.

states as explanations of behavior. However, learning theories, particularly *operant conditioning*, lend themselves well to discussions of motivation.

Operant conditioning theory states that human beings are motivated to engage in behaviors that are rewarded by the environment and to avoid those behaviors that are punished by the environment. Motivation itself influences whether or not the environmental contingency is rewarding or punishing or neutral. For example, if you are hungry, you are motivated to behave in ways that will bring you food, such as shopping or cooking. Hunger is the motivation and the environmental contingency is the food. The hunger, however, determines the value of the contingency. If you are not hungry, the food provides, at best, a neutral reward value. If you are starving, the food provides a rewarding experience.

Learning theorists also make a distinction between primary and secondary drives. Primary drives are those that lead us to satisfy a biological need, such as hunger or thirst. Secondary drives are those that have a learned reward value, such as any behavior that does not satisfy a biological drive. Through *chaining*, we link together secondary drives in order to fulfill a primary drive. Thus, all of the behaviors involved in the process of grocery shopping and cooking are a chain of satisfying secondary drives, which leads to the satisfaction of a primary drive by eating. This is how learning theorists explain how people are motivated by money and fame and power and other things that do not seem to have a direct biological basis.

Cognitive Approach

Several cognitive theories attempt to explain human motivation, each with a different focus. What they share is the notion of the motivated person as an active agent, deliberately working toward various goals, based either on desires or conscious choice, or both.

The *self-determination* theory focuses on what influences intrinsic motivation. Intrinsic motivation involves performing a behavior for its own enjoyment. Self-determination theory argues that if our innate needs of competence, autonomy, and relatedness to others are met, then our intrinsic motivation increases. If any of these needs are compromised, our intrinsic motivation decreases. Intrinsic motivation is also influenced by external rewards. Interestingly, there is some evidence that, under specific circumstances, the greater the external rewards, the less likely you are to be intrinsically motivated to perform the behavior. For example, paying a child for every A or B she receives in school may encourage studying, but, by itself, may not enhance her enjoyment of learning.

Other cognitive theories focus on our goal-setting behaviors. These theorists argue that it is our conscious goals that primarily motivate behavior. These goals are the outcomes that we desire, and are typically learned through social expectation. For example, we may set a goal of earning an A in a class because getting good grades is expected of us. Or, we may set a goal of selling more insurance policies because it is what the boss expects of us. Improvement in our goal-directed behavior involves the following factors:

- **Define attainable goal**   We must have a specific, attainable goal in mind. It should not be too difficult as to be unattainable, but also not so easy as to decrease motivation.

- **Experience discrepancy**   We must experience a discrepancy between what we have and what we want.

- **Receive feedback**   We must be able to gauge our progress toward the goal.

- **Commitment**   We must be committed to achieving the goal.

Humanistic Approach

The humanistic approach to motivation is largely based on the work of Abraham Maslow. He argued that our needs can be organized in hierarchical form, in which the lower level needs must be satisfied before we can attend to the higher level needs. At the lowest level of the hierarchy are our physiological needs, such as hunger and thirst. Once these needs are consistently met, we turn our focus to our safety needs. These include having adequate shelter, money, and protection. Once we have provided for these, we focus on our need for love and belongingness, or our need to be close to others. Next, we focus on our esteem needs, such as respect from friends and peers. Finally, the highest level of need is our need for self-actualization. This involves our need to express ourselves and grow as people. This can include the creative or performance arts, service to the community, expanding our skills, and the like.

Over the decades, Maslow's ideas have been investigated empirically with mixed results. The dominance of one type of need over another does not appear to be very strict. There are important cultural differences in the importance of various types of needs. It is also hard to explain things like working through lunch, not to mention self-sacrifice in war, in terms of Maslow's hierarchy.

Evolutionary Approach

The key to understanding the evolutionary approach to human motivation is to understand an underlying key concept of contemporary comparative psychology. Evolutionary theorists argue that all psychological attributes, including motivation, have been favored by natural selection to maximize reproductive success. Thus, we are motivated to engage in behaviors that increase the likelihood that our genetic material will survive and propagate to the next generation. For behaviors related to hunger, thirst, protection, and sex, this concept is obvious. Those people who do not have access to food will starve to death, and their genetic material along with them. Those who do not engage in behaviors that lead to both sex and conception will not pass their genes along to the next generation.

However, human behavior is more complicated than finding food, drink, and sex. For example, why do we bother to protect our children? After all, once we've passed along our genes, why put further resources into the next generation? The evolutionary approach would argue that we do this to maximize inclusive fitness. This means that because our child shares 50 percent of our genes, we are motivated to maximize the reproductive success of that child because we want those genes to continue. If you also protect (or feed or educate, etc.) your nephew, you are helping to pass along the 25 percent of the genes that you share. Thus, we are motivated to expend resources on those who share our genetic material. It is important to note that comparative psychologists do not argue that all behavior conforms to this thinking. However, there is good reason to think that behaviors related to hunger, thirst, competition for mates, mate selection, and so on, have such an evolutionary underpinning.

Variant View: The Panglossian Problem

It is important to understand that evolutionary arguments can be made for all sorts of motivations. For example, psychosocial motivations such as belonging and achievement might be explained by the fact that, as relatively small and slow primates lacking fangs and claws, we survive best in groups, and the fact that, as highly inventive and intelligent creatures, we survive better when some members of the group work hard to develop their skills. Philosophers have criticized the evolutionary approach because it can be used to explain almost anything. They call this the Panglossian problem.

Biological Motives

Certain motivations are considered more basic for two related reasons: Their connection to the survival of the organism, or to the survival of the species, is obvious. In addition, the connections to well-known basic biological activities tied to survival, such as feeding and sex, are also obvious.

HUNGER

Hunger is considered to be a basic motivational drive that leads to a variety of behaviors related, of course, to eating. On a basic level, we eat because we are hungry and we are hungry because we lack food. However, hunger is not quite that simple. There are many complex factors related to eating behavior, both biological and environmental.

Biological Factors

Modern theories of hunger focus on the influence of the brain, hormones, and blood sugar levels:

- **The brain**   Early research focused on finding anatomical structures in the brain. Various areas of the hypothalamus were believed to control the ability to regulate hunger. However, instead of brain structures, modern hunger theorists search for neural circuits within the hypothalamus.

- **Hormones**   Insulin and leptin are two hormones that contribute to the regulation of hunger. Insulin is secreted by the pancreas and extracts glucose from the bloodstream. Increased levels of insulin are associated with increased hunger. Leptin is produced by fat cells and released into the bloodstream, where it provides the hypothalamus with information about our fat stores. Increased levels of leptin tend to increase hunger. Thus, the more body fat one has, the more likely we are to be hungry.

- **Blood sugar levels**   Food taken into the body is converted into glucose, a simple sugar that provides energy. Early theories of hunger posited that increasing glucose levels decreases hunger, and decreasing those levels increases hunger. These levels are monitored by glucostats, or neurons in the body that are sensitive to glucose. Modern theorists argue that glucostats in the liver also play a role in regulating hunger, as they send signals to the hypothalamus.

It is important to note that the complexity of hunger regulation is not yet completely understood. It is believed that the interaction of the brain, hormone levels, and glucose levels in the bloodstream contribute to when we feel hungry and when we do not.

Environmental Factors

Biological factors tell us when we're hungry, but they do not necessarily regulate when we eat. We often forego food when we're hungry and eat when we are not. We also tend to find some types of food appetizing, but not others. There are several environmental and social factors that are also involved in determining our eating behavior. These include our learned habits and preferences and environmental food-related cues.

Much of what and when we eat is based on learned cultural preferences. Different cultures have developed food preferences over long periods of time based on what was available in their area. Grilled iguana sound appealing? If you were raised in certain parts of Mexico, it just might. Live beating cobra heart? Some Vietnamese consider it a delicacy (it makes you strong!). Cultural food preferences have evolved from a time when the food supply was hunted and gathered locally. We ate what we could find and/or kill. As human communities evolved into mobile societies, we retained our regional preferences, but we were also exposed to food from other cultures. Chinese cuisine is popular in the United States; French cuisine has influenced cultural preferences worldwide; and US fast food, unfortunately, can be found just about anywhere.

Although cultural food preferences influence what food we are exposed to, we also have individual food preferences. Why do some people prefer to eat seafood and others do not? Most individual food preferences are also based on learning. We may learn to enjoy a particular food item because we associated it with the experience of having eaten it in pleasant company, while in a good mood, or in conjunction with some other pleasant stimulus. Likewise, we may learn to dislike a food item because we associate it with a negative event, such as nausea or vomiting.

Finally, we can learn food preferences through social (or observational) learning. We can observe our dining companion enjoying the lobster and be more likely to try it ourselves. We can also observe (or hear about) a friend hating mushrooms and be less inclined to try those. Thus, we look at the contingencies others receive and either are inclined or disinclined to try the food item. Indeed, the easiest way to persuade a child to try a new food is to have an adult try it first, then make "yummy" noises to show a positive reward contingency.

There are also cues in the environment that signal what and when we eat. Food-related advertising is based on this principle. If something looks delicious, we are more likely to try it, and try it sooner rather than later. For example, did you ever notice that the hamburgers in television advertisements tend to glisten? Part of the reason is that humans have a genetic preference for high fat, and thus highly flavorful, foods. Early in our evolution, this helped us survive. The glistening in the ad is meant to suggest the fattiness, and thus the tastiness, of the burger. We also use the smell of food as a cue. Did you ever walk by a bakery and smell the baking bread? You may not have been hungry in the least, but now you're starving once you encounter the smell.

Finally, the situation can act as a food-related cue. If you make a habit of watching *The Daily Show* while munching on a bag of chips, then the show (including any reruns) will act as a cue for eating chips. The trick for overcoming the power of environmental cues is to stop and think about whether or not you are really hungry.

SEX

Sexual behavior, and all it entails, is a complex physical and psychological motivation. Generally, we engage in sexual behavior for two reasons: to reproduce and for pleasure. While both reasons have obvious physiological roots, they also have less obvious psychological influences.

Physiological Response

Masters and Johnson were the first to study the human sex response from a physiological point of view. Their landmark research yielded four phases of physiological sexual response:

- **Excitement phase**   The initial phase of sexual behavior in which a person becomes aroused at an accelerated rate.

- **Plateau phase**   Arousal continues to build to peak levels, but the rate slows.

- **Orgasm phase**   The peak of sexual excitement, which is accompanied by either ejaculation (in men) or muscle contraction (in women).

- **Resolution phase**   The physiological changes due to sexual arousal begin to subside.

We know, however, that there is more to sexual behavior than the physiological response. Why do we choose the sexual partners we choose? Are there different patterns of sexual behavior for men and women? What purpose does jealousy serve? We will examine these questions in terms of evolutionary psychology.

Evolutionary Influences

Given that the crux of evolutionary theory involves natural selection based on variations in reproductive success, evolutionary psychologists spend a great deal of time studying sexual behavior in many species, including humans. *Parental investment theory* states that mating patterns depend on the level of investment each sex contributes to producing and nurturing offspring, in terms of time, energy, and resources. The sex that makes the larger investment tends to be more selective when choosing partners and the sex that makes the smaller investment tends to compete for mating opportunities. In humans, females make a larger investment than men, even if both contribute equally to childrearing. Men only have to contribute viable sperm, but women must carry out gestation for nine months before going through the painful process of giving birth.

This helps explain the gender differences found in several areas of human sexual behavior. Overall, men tend to have more sexual partners than do women. Evolutionarily, this makes sense. Women have a proportionally larger investment in childbearing than do men. Because of this investment, women have a natural cap on how many children they can bear regardless of the number of sexual partners. Men have little investment in reproduction outside of copulation, so they are encouraged to mate with as many females as possible. Thus, while men must compete with other men for female sexual partners, women can be more discriminating in choosing a mate.

This theory also helps to explain the disparity in commitment willingness between men and women. Females tend to require a level of commitment in mating partners because it takes a great deal of resources (food, protection, etc.) to bear and raise children, thus assuring their child has a good chance at reproductive success. Males, on the other hand, tend to avoid long-term mating commitments because their "job" is to mate with as many females as possible, thus passing along their genetic material to as many offspring as possible. Some theorists argue that males would not commit at all if potential female mating partners did not require it.

Critical Caution

No evolutionary psychologist would argue that all human males avoid long-term mating commitment and all human females require it, or that all women care more about emotional infidelity and all men care more about sexual infidelity. They use the theory to explain why, on the whole, men and women tend to differ on these factors. Just as there was variation in our evolutionary past, there is variation in our evolutionary present. Natural selection is still at work. It takes a combination of genetics and environment to determine human behavior.

Given that there is a disparity in commitment expectations between men and women, we would also expect there to be a difference in jealousy. Indeed, men and women tend to be jealous over different things. Research suggests that men tend to be more bothered by sexual infidelity and women tend to be more bothered by emotional infidelity. Again, according to evolutionary theory, this makes sense. If a male is going to commit to a female and devote time and resources to childrearing, then he must be certain that any child she bears belongs to him. It makes little genetic sense to invest resources in a child who does not share your genes. Thus, males tend to be alert to possible sexual infidelity. Females, on the other hand, tend to be more alert to emotional infidelity. The time, energy, and resources a male invests in the female and her children are rooted in emotional commitment. Emotional involvement with another woman could signal the loss of the male's investment in her children.

Finally, men and women tend to differ on the attributes they seek in a potential mate. The parental investment theory predicts that men should seek women with good reproductive potential and women should seek men who can protect and provide. These were humans' needs in our evolutionary past. Do they still hold? A large-scale study conducted in 37 countries provides evidence that it does. Men tend to value physical attractiveness and youth, which signal good reproductive potential. Women tend to value intelligence, social status, and ambition, which signal the ability to provide resources and protection. Even more interestingly, these results held across cultures, including capitalist societies, third-world cultures, and a vast array of economic systems. Thus, across a vast array of environmental variety, human mating preferences tend to conform to evolutionary theory.

Sexual Orientation

A discussion of human sexual behavior would not be complete without a discussion of sexual orientation. Sexual orientation refers to a person's emotional, physical, and sexual attraction to the opposite sex (heterosexuality), the same sex (homosexuality), or to both (bisexuality). In reality, people are rarely exclusively heterosexual or homosexual. Our sexual orientation tends to fall on a continuum. People who consider themselves exclusively heterosexual frequently report that they have had homosexual experiences and fantasies, and those who consider themselves exclusively homosexual also report heterosexual experiences and fantasies. Thus, even sexual orientation is not as clear cut as we have been lead to believe outside of science.

The scientific community is just beginning to study sexual orientation, but we have already accumulated a good body of research that indicates that sexual orientation has strong genetic and biological influences. Indeed, attempts to explain sexual orientation via psychoanalysis (strong mother and weak father) and learning theory (positive associations with the sex of choice) have failed to support these theories. Thus, we will focus on the biological and genetic factors that have empirical support.

The early biological/genetic studies of sexual orientation examined the prevalence of homosexuality in our own and other species. Although exclusive homosexuality is rare in other species (as it is in humans), homosexual behaviors are frequently observed. Thus, there is ample support for the notion that homosexuality is not exclusive to humans. We also have evidence that homosexuality occurred in ancient cultures, such as Rome and Greece, and still occurs in almost all cultures across the world. In a large part of the world ranging from Melanesia (the islands off the northern coast of Australia, including Fiji, New Guinea, and the Solomon Islands) to Sumatra (Indonesia), most males participate in homosexual activities during the years before marriageable age, and the behavior is considered completely normal for that period of life. Thus, homosexuality has existed throughout human history and still exists in almost every culture today. Each of these pieces of evidence suggests that sexual orientation has a genetic or biological component.

Further evidence can be found when examining differences in the neural anatomy between heterosexual and homosexual men. In a landmark study, LeVay found that a specific set of nuclei in the hypothalamus, known to influence sexual behavior, was twice as large in heterosexual men as it was in homosexual men and heterosexual women. In animal studies, the surgical disruption of this nuclei leads to atypical sexual behavior. Thus, preliminary evidence suggests that brain organization and anatomy also influence sexual orientation.

Finally, heritability studies support the hypothesis that sexual orientation is highly heritable in both men and women. Several studies have shown that male relatives of male homosexuals have a much higher rate of homosexuality than would be expected. While homosexuality rates in the general population run between 4 percent and 7 percent, brothers of male homosexuals have a 25 percent rate of homosexuality. When comparing identical, fraternal, and adopted siblings of homosexual men and women, similar results were found. Homosexuality rates for identical twins were over 50 perecent, approximately 20 percent for fraternal twins, and roughly 10 percent for adopted siblings. Geneticists have also been able to create homosexuality in fruit flies by producing genetic mutations. These results taken together strongly support the claim that sexual orientation has a genetic component.

Psychosocial Motives

Motivations for which both the biological and evolutionary basis is less clear are often considered separately. Two of these are the need to belong and the need to achieve.

BELONGING

The vast majority of primate species are gregarious, rather than solitary, animals. We humans are no exception. We have evolved to survive by working together and protecting one another. Primates share the responsibilities for protecting their young; for defending against predators; and for gathering, hunting, and scavenging food. Spending time with other people has benefits to both survival and procreation. We descendants of all of those other monkeys are motivated to spend time with one another.

Unlike food, drink, and sex, the biological correlates of being with other people are not well-understood. Nonetheless, in evolutionary terms, our ancestors have needed one another long enough for companionship to have evolved into a primary reinforcer. We may be hardwired to seek out the sight and touch and perhaps even the smell of other human beings.

In contemporary society, human groupings are both beneficial and detrimental to society as a whole. Government, education, sex, charity, and many other aspects of society derive their benefits due to people's operating in groups. The specialization of modern civilization, where people become doctors, teachers, police officers, soldiers, accountants, nurses, and so on, so that each of us does not have to know everything about how to survive, is all about belonging to the group. On the other hand, the motivation to join groups also results in gangs and mobs and the acceptance of authoritarian rule. People work together in both positive and negative pursuits.

Psychologists have identified at least three different types of belonging motivation: *intimacy*, *affiliation*, and *attachment*. Intimacy needs are satisfied by spending time with others with whom we share private information, about whom we care, and toward whom we have warm feelings. Affiliation needs are satisfied by spending time with those with whom we share good news about our own lives, general news about others, and engagement in group activities. Attachment needs (which were discussed in more detail in Chapter 3,"Developmental Psychology") involve being physically close to those who provide us with protection and/or support in times of need.

Big Background: Why We Want to Belong

The motivation to belong may have a great deal to do with various phenomena discussed in Chapter 2, "Social Psychology." Conformity, obedience, social comparison, and other behavioral patterns that help maintain social cohesion may themselves be maintained by our need to belong.

People differ with respect to how much they need intimacy and affiliation. Irrespective of these differences, when these needs are not satisfied, bad things happen. Depression, poor physical health, stress, and shortened lifespan are all associated with unsatisfied social needs. The sharing available in intimate relationships seems to prevent depression following a crisis. Infants deprived of attachment can even die as a result and may suffer lifelong difficulties if they live.

Recent research shows the importance of attachment needs even in adulthood. When we feel that we will be supported, we are more likely to be compassionate and generous toward others. Behavior such as altruism and care-giving may be more common when people have their attachment needs satisfied. If we work to address these individual needs, we may improve the global situation with respect to societal ills such as poverty, human rights violations, even war.

ACHIEVEMENT

Another psychosocial need that has been identified by psychologists is the need for achievement. Achievement is harder to define than belonging. Achievement means success, but success in whose eyes? Psychologists distinguish between people who seek achievement on their own terms (called *mastery*), and those who seek achievement measured by the approval of others (called *social achievement*). (Note that *mastery* here has a very different meaning than it does in psychological testing.)

In addition, there is the question as to what sorts of things are deemed to be important achievements. In individualistic societies, including the United States, achievement is usually measured in terms of individual accomplishments, whether measured in terms of benefits to oneself or in terms of identifiable contributions to the good of others or society as a whole. In more collectivist societies, such as China, achievement is measured in terms of success in promoting the work of the group, whether it is family or friends or society as a whole.

Empirically, psychologists have found various things correlate with high scores for achievement motivation. People with strong achievement motivation are more willing to delay gratification. They are more persistent in their endeavors. When they experience failure, they are more likely to attribute the failure to bad luck than to their own abilities. This may be part of why they go back and try again.

Intriguingly, high achievers seek tasks that are neither too easy nor too hard. Presumably, while easy tasks are most likely to produce short-term success, they do not provide the opportunity to learn and improve one's performance so as to better succeed in the long-term. Tasks that are too hard are too likely to result in failure.

Variant View: Habits and Motivation

An important difficulty in studying achievement motivation is the reliance on questionnaires to determine how much the subjects are motivated to achieve. These tests consist of questions about how much the subject wants to succeed at this or that. For the most part, these tests ask about individual achievements, because they are authored by Western psychologists with Western values and Western views of achievement. Highly successful folks in collectivist societies do not rank as high on these tests as do Westerners.

This raises a question: Persistent people may succeed more. So may people who attribute their failures to bad luck, who delay gratification, and who choose tasks of intermediate difficulty as their skills improve. This success may lead to a desire for more. High scores for achievement motivation may just be the effect of having the sorts of habits that lead to success as defined by one's own society.

Applications

The study of motivation may not be tied closely to studies of how people function. The relationships between the different types of motivation may not be fixed. The results of many motivation studies may be culturally dependent. Even the definitions of some kinds of motivation may be culturally dependent. But none of this means that the study of motivation is unimportant. Determining who can achieve what and how to maximize the achievements of different kinds of people is vitally important to society. It is in the area of applied research where the study of motivation is most important.

Two areas of society where motivation is most important are business and the military. In a highly individualistic society like America, institutions evolve to take advantage of individual motivation. The team sports most popular in America are those where individual stars can make the entire team a success. (This is one reason why soccer, the most popular sport in the world, is not so popular here. The most important contributor to a soccer team's success is the goalie, and he almost never does any scoring.) Education focuses on individual achievement and rewards motivated individuals. Government is organized around leaders. However, neither commerce nor warfare can be conducted successfully by relying on individuals. In order to make these collective enterprises work, applied psychologists have done extensive research into how to motivate individuals to work effectively in a group.

WORK

Applications of psychology to the business world are an example of the challenging problems and special rewards of attempting to apply psychological science to the real world. As such, we will discuss them briefly here. *Industrial/organizational (I/O) psychology* is the name given to this sort of research. The psychology of motivation has an important role in business because more highly motivated workers (in some sense of that term) mean higher productivity and higher productivity contributes to higher profits.

I/O psychology focuses both on people and situations. *Personnel psychology* (now more commonly called *Human Resources (HR) psychology*) is concerned with how to identify the people who will do best at a particular job and how to get the best out of workers already on the job. *Organizational psychology* is concerned with how to structure the job and the company so as to get the best out of everyone. *Human factors psychology* addresses the points of contact between the worker and the job. For example, designing comfortable chairs (a part of ergonomics) and user-friendly software are both issues in human factors.

The challenges of I/O psychology are very typical of the challenges faced in all sorts of applied psychology. In America, business is atypically focused on the near term, at least as compared with businesses in other parts of the world. American businesses are often unwilling to expend money now to identify current problems and find solutions that will improve the bottom line in the future, particularly if the future is more than a year away. Applied research requires investment, not only of money, but of time and personnel and other resources.

In addition, the demands of scientific methodology do not always fit in nicely with the ongoing requirements of running a business from day-to-day. For example, suppose we are developing a test that will help identify which applicants are best suited to a particular job. From the perspective of psychological testing, the best strategy is to test a large number of applicants, hire them all, and see who does best. Of course, this means hiring a lot of clearly unqualified people, which would not only be costly, but very likely damaging to the business.

LEADERSHIP

An intriguing part of organizational psychology is the study of leadership. Bosses are people and thus a proper topic for psychological study. In terms of worker productivity, however, bosses are an important part of the workers' environment. Thus, in business, the study of leadership focuses on which behaviors on the part of the leader promote better behavior on the part of the workers. In terms of the subject of this chapter, how can leaders best motivate workers?

In the military, the focus is somewhat different. While managers and supervisors in business have many duties, the principal responsibility of line officers in the military is to lead. The military works hard to identify the personal characteristics of officers that identify them as being the best potential leaders. In other words, business tries to modify the behavior of the leaders they have, while the military tries to hire and promote folks who will be the best leaders.

Summary

The study of motivation began as an attempt to understand our common sense views of the causes of behavior in scientific terms. These attempts to integrate motivation into larger theoretical schemes, such as psychoanalysis or behaviorism, have not received much attention in recent years. Instead, the study of motivation has focused on the biophysical for individual motivations and on potential applications for social motivations. In both cases, the study of motivation has become a separate field of study in psychology.

Quiz

1. The hallmark of the _____ approach to motivation is the notion that basic drives can exert their influence unconsciously.

    (a) Behavioral

    (b) Cognitive

    (c) Humanistic

    (d) Psychodynamic

2. Learning theorists use _____ to explain how people are motivated by money and fame and other things that do not seem to have a direct biological basis.

    (a) Primary drive

    (b) Chaining

    (c) Secondary drive

    (d) None of the above

3. According to self-determinism theory, intrinsic motivation is tied to our need for _____.

   (a) Competence

   (b) Autonomy

   (c) Relatedness

   (d) All of the above

4. According to Maslow, which of our needs is the highest level of the hierarchy of needs?

   (a) Physiological needs

   (b) Safety needs

   (c) Self-actualization

   (d) Esteem needs

5. We are motivated to expend resources on those who share our genetic material to maximize _____.

   (a) Inclusive fitness

   (b) Safety needs

   (c) Autonomy

   (d) All of the above

6. Modern biological theories of hunger focus on the influence of which areas of research?

   (a) The brain

   (b) Hormones

   (c) Blood sugar levels

   (d) All of the above

7. According to the parental investment theory, the sex that makes the _____ investment in producing and nurturing offspring tends to be more selective when choosing partners.

   (a) Smaller

   (b) Equal

   (c) Larger

   (d) Solo

8. Generally, men tend to become jealous over _____ infidelity, and women become jealous over _____ infidelity.

   (a) Sexual, sexual

   (b) Sexual, emotional

   (c) Emotional, sexual

   (d) Emotional, emotional

9. Which of the following is not a type of belonging motivation?

   (a) Intimacy

   (b) Achievement

   (c) Affiliation

   (d) Attachment

10. Which of the following areas of research is important to industrial/organizational psychology?

    (a) Personnel psychology

    (b) Organizational psychology

    (c) Human factors psychology

    (d) All of the above

CHAPTER 11

Learning

Animals, including humans, not only act in and on the world, they change their patterns of acting in order to act more effectively. These changes are what we call learning. Psychologists are interested in the systematic ways that behavior changes to cope with novel circumstances.

Overview

Learning is an important area of study because it impacts almost everything we do. Through learning, we not only acquire academic skills, such as reading and math, we also acquire the knowledge we need for everyday functioning. When you master the techniques of driving, cooking, or finding your way around campus, you're engaged in forms of learning. It is also important to realize that learning does not just happen during childhood. We are learning constantly throughout our lives. You just learned something by reading this paragraph.

The subject of learning poses a particular problem for psychologists. It is not as easy to determine when learning has occurred as you may think. It is not sufficient

Topical Talk: Studying Nonhuman Animals

More than almost any other area of psychology, what we know about learning has been derived from experiments with nonhuman animals. This raises the question as to how much we can or should rely upon data from nonhuman animals in understanding the psychology of humans.

On the positive side, the enormous genetic, physiological, and neurological similarities between humans and other animals can give us confidence that much of the basic apparatus for various psychological functions are the same for other animals as they are for us. In addition, the gloss of civilization and language means that human subjects can interact with and interfere with experimenters in ways that other animals cannot. Finally, since all psychologists are human, they have biases toward their fellow humans that may not be present for other sorts of animals.

On the negative side, psychological functions are governed mostly by the brain and the human brain appears to be a good deal larger (proportionately, at least) and somewhat more intricate than the brains of most other animals. Many (but not all) of the characteristics that make humans unique, such as language and civilization and art and music and science (and becoming psychologists), seem to be dependent more upon the brain than upon anything else. (As Mark Twain put it, "Man is the only animal that blushes—or needs to.") Finally, even though animals seem simpler than we do (making human psychology just animal-psychology-plus), in reality, almost every animal has its own uniquely complex functions that far exceed the capacities of the human equivalent. For example, humans rely first and foremost on vision to understand the world and cope with it. Rats use smell. The brain of the rat is much smaller than ours, but its ability to deal with smells far exceeds our own. This leads to an interesting research issue: If we wish to use the rat to study how the senses allow humans to maneuver about in the world, should we study rat vision or rat smell?

to just ask the student if she has learned something. A person may not be certain whether or not he has learned something, or he may lie because he does not want to look bad. Psychologists are also interested in animal and infant learning, where the subjects are not capable of verbalizing whether or not they have learned.

When we cannot ask the subjects what they have learned, we must examine changes in their behavior. But simple change is not enough. Behavior changes due to development, aging, injury, and reactions to various conditions such as heat and cold or hunger and thirst, etc. Learning is adaptive change due to experience. As we will see later on in the section on Cognitive Learning, some contemporary researchers define learning so as to exclude these other sorts of behavioral changes. For now, we will look at how learning is studied to see how other sorts of behavioral changes are excluded experimentally.

Big Background: Victorian Excitement

During Victorian times, there was huge popular interest in animal intelligence. Stories and books and newspaper articles abounded with anecdotes about pets finding their way home across long distances, dogs who learned to unlatch gates in order to open them, horses who did arithmetic, and even apes raised as people in human homes. This fascination with the abilities of nonhuman animals derived from the enormous excitement generated by the work of Charles Darwin. Darwin's assertion that we humans are related to the other animals led to a fascination with the mental aptitudes of our distant relations.

HISTORY

The first psychologists to study learning were interested in what they called *animal intelligence.* By this they meant, how are nonhuman animals clever in ways that are similar to how humans are clever? To what degree are other animals capable of those things we think of as distinctly human? Dogs and cats and other pets, and even wild animals, were reported to perform remarkable tasks. Beginning in the 1870s, these early psychologists tried to find out if these animals were really smart, or just lucky.

Two of these researchers, Ivan Pavlov and E. L. Thorndike, who were heavily influenced by the physiological researchers of the time, moved these studies into the controlled conditions of the laboratory. Pavlov began by looking at *reflexes,* fixed behaviors that reliably occurred after specific environmental events called *stimuli.* He showed how animals could learn to produce similar behaviors under different conditions. He called this the Law of Reinforcement. (It is now called the *Law of Contiguity.*) Thorndike attempted to replicate the stories of clever pets and ended up demonstrating how animals could learn to produce novel behaviors. He called this the *Law of Effect.*

The work of these researchers and others inspired an entire movement within psychology: behaviorism. Behaviorism was not just a methodological approach; it was also a philosophy of how to create a scientific psychology.

Behaviorists believed that if psychology was going to be a scientific discipline, then it must study only behaviors that can be directly observed. Along this line, they made three important assumptions. First, behavior is determined in some way and thus should follow certain laws. Behaviorists argued that the job of scientific psychology is to discover these laws. Second, the explanations of behavior based on internal causes and mental states are useless for a scientific psychology. Behaviorism focuses on behaviors that are observable and look toward the lawfulness of behavior for an explanation. Finally, the environment shapes behavior.

Big Background: Associationism

Because psychological stimuli are not metal prods and the responses we study are more than simple muscular reactions, behaviorists looked to philosophy for a more abstract description of behavior than the simple reflex. What philosophers call an *association* is a general relationship between environmental circumstances and action, where the circumstances (stimuli) are said to cause the action (responses). Many psychologists who are not behaviorists are still associationists in a broad sense.

The physiologists who influenced the early behaviorists studied reflexes by prodding the nerves and measuring muscular reactions. Early behaviorists searched for lawful relationships between stimuli in the environment and the response of the organism. In other words, they attempted to reduce all behavior to complex interactions between various reflexes.

Beginning in the mid-1930s, three behaviorists, Hull, Tolman, and Skinner, dominated psychology. Hull and Tolman expanded on Pavlov's work and kept within the framework of the stimulus-response approach. Skinner expanded on Thorndike's work and gradually abandoned the stimulus-response approach.

CURRENT APPROACHES

In the 1950s and 1960s, the cognitive approach overthrew behaviorism's dominance in psychology. Early cognitivists challenged behaviorist assumptions, but focused their own research efforts elsewhere. Cognitive learning theorists such as Bandura made important contributions by identifying types of learning not explainable in terms of either Pavlov's Law of Reinforcement or Thorndike's Law of Effect.

Beginning in the late 1970s, a group of cognitivists, frustrated with cognitivism's failure to explain basic mechanisms of learning, developed a computer simulation approach to studying learning called *connectionism*. Their computer models resemble groups of neurons and are called *neural networks*.

Currently, there are three major American approaches to the study of learning. Followers of B. F. Skinner's approach, called *behavior analysts*, represent the last remaining school of behaviorism, called *radical behaviorism*. Tolman's focus on expectations in learning inspired contemporary cognitive approaches to learning. Today, Hull's influence is not strong. Indirectly, however, Hull's approach led to work that eventually resulted in mathematical and computational approaches, including connectionism.

A fourth approach, much more popular in Europe than in America, is *ethology*. Based on work with various nonhuman animals in naturalistic settings in the 1950s

by biologists such as Lorenz and Tinbergen, ethologists focus on species-specific learning outside the laboratory. Ethology represents an important critique of the notion shared by both behaviorists and cognitivists that one general set of laws governs all of learning.

Conditioning and Animal Learning

The first scientific research on learning used the methods of *conditioning*, arranging circumstances for an animal (including some human animals) so that learning how to cope with those circumstances will benefit the animal. Conditioning relies upon an animal's ability to adapt its behavior to take advantage of circumstance in order to determine how specific situations create opportunities for specific kinds of learning. The logical structure of these situations is called *contingencies*.

Traditionally, conditioning is separated into two types, *classical conditioning* and *operant conditioning*. These two types of conditioning developed separately in two closely related experimental traditions. Classical conditioning involves causing the animal to exhibit old behaviors in new situations. Operant conditioning involves causing the animal to exhibit new behaviors at specific times when it can be beneficial.

CLASSICAL CONDITIONING

Classical conditioning is a method for producing learning based on the work of Ivan Pavlov. Pavlov was a trained physiologist who stumbled on this phenomenon while studying canine digestion. He was investigating how the mouth prepares for food by secreting saliva and found that the mouth also secretes saliva when food is smelled or seen. Because this phenomenon was not based on a physiological stimulus—food actually being placed in the mouth—Pavlov assumed that this was a psychological process. This discovery laid the foundation for Pavlov's subsequent investigations into classical, or *Pavlovian*, conditioning.

Critical Caution

The study of conditioning involves the use of a great deal of specialized vocabulary. Although it can be difficult to keep track, it is important to use the terminology correctly. For example, *conditioning* is what the experimenter does to the subject animal and *learning* is what the subject animal does as a result. (But don't forget that we experimenters are animals, too.)

Basic Definitions

The terminology and definitions in classical conditioning are a source of confusion for many introductory psychology students. The key to understanding the terminology is to define two sets of definitions, *stimulus vs. response* and *conditioned vs. unconditioned*. A stimulus is the thing that causes a potential response. A response is the behavior caused by a stimulus. (The terms *stimulus* and *response* were taken from physiology, but have a much broader meaning in psychology.) The distinction between conditioned and unconditioned is simply the distinction between learned and automatic. By combining these two sets of definitions, we arrive at the following terms: *unconditioned stimulus, unconditioned response, conditioned stimulus*, and *conditioned response*.

Basic Methodology

The experimental methodology described here is based on Pavlov's initial experiments. There are many variations on this methodology, such as the stimulus used, the response elicited, and the organism being studied. By using the basic experimental methodology, we can identify the four basic terms just described. Pavlov placed a dog in an apparatus that restrained its movement while allowing the dog to be near a food dish. A tube was inserted into the dog's cheek so that saliva flowed from the salivary glands into a container. The experiment began by presenting the dog with a neutral stimulus, such as a bell. The stimulus is neutral because the bell does not automatically cause the dog to salivate. Several seconds after the bell was sounded, food was dropped into the dog's dish. When the dog ate the food, the dog salivated. This procedure was repeated until the bell alone began to elicit salivation. Therefore, the sound of the bell is paired with the presentation of food until the bell alone causes salivation.

Fun Facts

The terms *conditioned* and *conditioning* in the psychology of learning are actually due to a mistranslation of Pavlov's original work into English. In Russian, Pavlov actually spoke of "conditional" and "unconditional" reflexes. He was not speaking of the things he was doing to the animals. He was merely making a distinction between reflexes that were found unconditionally, when the animal was first tested in the laboratory, and those that only appeared conditionally, due to the training Pavlov was providing. Now we call that training *conditioning*.

Let's examine the basic definitions in terms of the experimental methodology:

- **Unconditioned stimulus**   The food in the mouth is the unconditioned stimulus. Food in the mouth is the thing that automatically elicits the salivation response without learning.

- **Unconditioned response**   Salivation to the food in the mouth is the unconditioned response. The dog does not have to learn to salivate when food is put in its mouth.

- **Conditioned stimulus**   The sound of the bell is the conditioned stimulus. It begins as a neutral stimulus and is repeatedly paired with the unconditioned stimulus to elicit the salivation response. Thus, while the sound of the bell did not automatically cause the dog to salivate (which is why it is called "neutral"), the dog learned to salivate to the sound of the bell because it was repeatedly paired with the food. The sound of the bell has been *conditioned* to elicit salivation.

- **Conditioned response**   The dog's salivation to the sound of the bell is the conditioned response. The dog learned to give a response to a previously neutral stimulus (the bell) because the stimulus was paired with another stimulus (the food) that automatically elicited the response. Salivation has been conditioned to the sound of the bell.

It is important to note that both the conditioned and unconditioned response involve salivation. The distinction is exactly which stimulus caused the salivation. It is unconditioned if the salivation is in response to the unconditioned stimulus (food in the mouth), and conditioned if in response to the conditioned stimulus (sound of the bell). (In the laboratory, the precise form of the conditioned salivation can be distinguished from the form of the unconditioned salivation.)

In order to establish the conditioned response, the conditioned stimulus needs to be repeatedly paired with the unconditioned stimulus. The strength of the conditioned response is determined by *how* the two are paired. *Forward conditioning* is the easiest procedure used to establish a strong relationship between the unconditioned and conditioned stimuli. Forward conditioning consists in presenting the conditioned stimulus a few seconds before presenting the unconditioned stimulus. In the experimental methodology previously described, the bell is rung a few seconds before the food is presented.

Backward conditioning is a less effective procedure. In backward conditioning, the conditioned stimulus is presented a few seconds after the unconditioned stimulus. The bell is rung a couple of seconds after the food is presented. A third pairing procedure involves randomly presenting either stimulus first. Thus, sometimes the bell is rung first and other times the food is presented first. There is no consistent

> ## Variant View: Perception in Pavlov
>
> The difference in the strength of the conditioned response due to the ordering of the stimuli also demonstrates a contention of many modern behaviorists. They argue that classical conditioning is more complex than Pavlov originally conceived. They stress the fact that one stimulus gives the organism information about the arrival of the other stimulus. Therefore the organism needs to perceive a relationship between the two. Thus, they claim that Pavlov's dogs perceived a connection between the sound of the bell and the food in that the sound of the bell signaled that the food was about to arrive. Postulating that a dog *perceived* something violates the assumption of behaviorism that prohibits considering internal or mental states as explanations of behavior.

relationship between the two stimuli. This results in either weak conditioning or no conditioning whatsoever.

OPERANT CONDITIONING

Operant conditioning (also called *instrumental conditioning*) is another type of procedure that produces learning. It is based on the effects of rewards and punishments on behavior. Essentially, it involves learning to change voluntary actions based on the consequences they bring.

Thorndike was the first to study the effects of reward and punishment on animal behavior in the laboratory. He constructed an apparatus, called the *puzzle box*, that required a specific combination of moves to open a door. Once the door was opened, the animal could escape and eat food placed outside the box. When he placed a hungry cat in the box, he found that the cat's behavior was erratic at first. The cat would scramble about and make the desired response only accidentally. However, with repeated trials, the cat gradually became more proficient at escaping until it could open the door immediately.

Thorndike concluded that the cat learned to escape because the escape responses were associated with a desirable consequence, food. From these studies, Thorndike postulated the Law of Effect, which states that responses that lead to unsatisfying consequences will be weakened and are unlikely to be repeated, whereas responses that lead to satisfying consequences will be strengthened and are likely to be repeated.

Skinner furthered this work by developing a wide variety of highly replicable procedures using enclosed chambers with specific devices, such as levers and buttons, that the animal could operate. Skinner called the levers and buttons and other devices, *operanda*. He called the measurable results, the lever presses and button pushes,

Tips on Terms: Reward Versus Reinforcement

Skinner avoided the term *reward*, preferring to use the term *reinforcement* when discussing stimuli that followed a response and increased its frequency. His concern was both that *reward* suggested positive feelings he could not be sure were happening and it suggested that the properties of the stimulus itself increased responding, rather than a relation between the stimulus and the individual. Chocolate is reinforcing for some children, but not for all. Rewards are what we give people in hopes they will be reinforcing.

Regrettably, Skinner did not choose a novel term to replace the term *punishment*. This creates confusion, because the word *punishment* has all of the same difficulties that *reward* does.

In our discussion of operant conditioning, we will use the standard technical terminology and avoid the use of the term *reward*.

---

operant behavior. Our discussion of operant conditioning will be based largely on Skinner's work.

Basic Definitions

As with classical conditioning, students generally have trouble keeping the basic definitions used in operant conditioning straight. The key here is to remember two distinctions; *positive vs. negative* and *reinforcement vs. punishment*. The first distinction involves the act of giving something versus taking something away. A consequence is positive if something is given to the learner and negative if something is taken away from the learner. The second distinction involves whether or not the learned behavior is supposed to increase or decrease in either occurrence or frequency. Reinforcement is a consequence that causes a behavior to be repeated or increase in frequency. Punishment is a consequence that causes a behavior to be suppressed or decrease in frequency.

Combining these two distinctions gives us the four basic definitions in operant conditioning:

- **Positive reinforcement**   The learner is presented with a pleasant consequence in order to cause the occurrence of or to increase the frequency of a behavior. For example, giving money to a child so that he will clean his room.

- **Negative reinforcement**   The learner has an unpleasant consequence removed in order to cause the occurrence of or to increase the frequency of a behavior. For example, taking away extra chores from a child so that she will clean her room.

Tips on Terms: Positive and Negative, Reinforcement and Punishment

The distinction between *positive* and *negative* consequences is not always clear. Providing food relieves hunger. Giving money removes debt. Taking away chores provides leisure time. Heat can serve to reinforce behavior in an animal that is too cold, but might serve to punish an animal that is already warm. When it is reinforcing, are we providing heat, or removing cold?

The distinction between *reinforcement* and *punishment* is not so ambiguous. When a narrowly defined behavior increases or decreases in frequency, it replaces or is replaced by a wide variety of other behaviors. This is why reinforcement is more effective than punishment. When the behavior of talking back to one's parents is decreased, it may be replaced by an even more problematic behavior, such as temper tantrums. Honey really does work better than vinegar.

- **Positive punishment**   The learner is presented with an unpleasant consequence in order to prevent the occurrence of or to decrease the frequency of a behavior. For example, spanking a child when he talks back to his parents.

- **Negative punishment**   The learner has a pleasant consequence removed in order to prevent the occurrence of or to decrease the frequency of a behavior. For example, taking away a child's privileges when she talks back to her parents.

Basic Methodology

The basic methodology described here is based on Skinner's general methodology. There is a great deal of variety to be found, but these generalizations can apply to almost any operant conditioning experiment.

Skinner constructed an operant chamber, also known as a *Skinner box*. It is a small compartment in which reinforcement or punishment is automatically delivered immediately after an animal—most often a rat or a pigeon—performs a specific target behavior. Typically, the box contains a food or water delivery system (called a feeder or dipper) that is connected to a lever (usually with a bar handle) or button that the hungry animal is to press. Thus, a *consequence* (food or water) is presented reliably almost immediately after the target behavior (the bar press) is detected. The bar is also connected to a cumulative recorder that produces a graph of the rate of bar presses.

Learning happens because the target behavior is followed by a programmed consequence; therefore, the first step is to teach the animal to perform the target behavior. For example, rats are curious animals by nature and will explore the operant chamber once they are placed in it. Sometimes, it is possible for the rat to stumble on the target response by sheer luck. They happen to press the bar when they are exploring and are reinforced with food. The problem with this procedure is that it can take quite a bit of time.

A more efficient procedure is to *shape* the response. Shaping is a process where the experimenter trains the animal to make the desired response by reinforcing closer approximations of the response until the response is made consistently. For example, the experimenter, equipped with a button that opens the feeder, will place a hungry rat in the operant chamber, observe the rat's behavior, and push the button whenever she sees the rat pointing his nose in the general direction of the bar. Once the rat is consistently facing the direction of the bar, the experimenter will reinforce it for exploring the area around the bar until it does this consistently. Next, the rat will be reinforced for sniffing the bar. Finally, the rat will begin pressing the bar and will be reinforced without the experimenter's intervention, because the bar is hooked up to the feeder.

In the basic operant procedure, there are only two elements, the *behavior* (responses) and its *consequences* (reinforcing or punishing stimuli). However, stimuli called *antecedents* (or discriminative stimuli) can also be presented before the behavior occurs. Similarly to the conditioned stimuli of classical conditioning, discriminative stimuli can inform the animal that the behavior will be followed by some consequence.

Reinforcement Schedules

Once the desired response has been established, various reinforcement patterns can be employed to maintain, increase, or decrease the rate of response. The key to maintaining the response is to keep up the reinforcement, but the *schedule of reinforcement* determines the rate of responding. By altering the contingency under which reinforcement is delivered, we can change both the response frequency and persistence.

The baseline schedule of reinforcement is *continuous reinforcement*. This schedule provides one reinforcement each time the behavior occurs. The graph of the response rate is used as a baseline for comparing changes to the continuous re inforcement schedule. Any change to the continuous reinforcement schedule results in a *partial reinforcement schedule*.

There are four basic types of partial reinforcement schedules, based on two sets of distinctions. The *fixed vs. variable* distinction is based on whether the animal is reinforced consistently within an experimental session or if the reinforcement varies.

The *ratio vs. interval* distinction is based on whether the animal is reinforced for a specific number of behaviors or after a certain amount of time. These are described next:

- **Fixed-ratio schedule**   Reinforcing the behavior after it occurs a specific number of times, where the number of times does not change within the experimental session. For example, during a one-hour session, the rat is reinforced after every 10 bar presses. This results in a rapid rate of responding with a short pause after reinforcement.

- **Fixed-interval schedule**   Reinforcing the behavior the first time it occurs after a specific interval of time has elapsed, where the length of time does not change during the session. For example, the rat is reinforced for the first bar press after every 30 seconds. Other responses are not reinforced. This results in a considerable pause immediately after reinforcement, with a moderate response rate after the specific interval has passed. In other words, during the early part of the 30-second interval, when responses have not been reinforced, response rate is very slow or entirely absent. Later on, closer to the 30-second point where responses are reinforced, response rate increases.

- **Variable-ratio schedule**   Reinforcement is given after an average number of responses that varies within the experimental session. For example, the rat is reinforced after 5 bar presses, then after 10 bar presses, then after 3 bar presses, and so on. The average might be one reinforcement for every five responses, but there is no way to predict how many responses are needed for any specific reinforcer. This results in a very high, steady rate of responding.

- **Variable-interval schedule**   Reinforcement is given after the first bar press after an average time period has elapsed within the experimental session. For example, the rat is reinforced after 10 seconds, then after 30 seconds, then after 5 seconds, and so on. The average might be 15 seconds, but there is no way of predicting how soon reinforcement will be available after the last reinforcement. This results in a lower steady rate of responding.

There are hundreds of variants on the basic reinforcement schedules listed here. They allow the operant researcher to investigate many types of behavioral phenomena under tightly controlled circumstances. An important feature of some of these schedules is the use of multiple operanda, levers, buttons, chains, and so on. Pressing one lever can be reinforced differently than pressing the other. Different stimuli can be placed behind different buttons. This allows experimental tests of choice between levers. Choice is also investigated by cognitive psychologists studying thinking. In this way, behavioral and cognitive tests of the same phenomena can be compared.

Variant View: Common Misconceptions About B. F. Skinner

B. F. Skinner (1904–1990) was one of the most important psychologists of the twentieth century, but also one of the least well-understood. Here is a list of common misconceptions about his views:

- *Skinner was a stimulus-response (S-R) psychologist.* False. As a key developer of behaviorism, Skinner made extensive use of the terms, *stimulus* and *response*. Eventually, he came to doubt that the environment could be neatly divided into stimuli and that behavior could be neatly divided into responses.

- *Skinner thought all behavior was caused by the environment.* Wrong on two counts. Skinner believed that the immediate environment (discriminative stimuli) played only a secondary role in producing behavior. It was the history of reinforcement (called the response-stimulus relation) that played the primary role. He also understood the vital importance of genetics.

- *Skinner did not believe thoughts or feelings or intentions or beliefs should be part of psychology.* Not quite. Skinner did not believe that thoughts or feelings or intentions or beliefs should be used as scientific *explanations* in psychology. For Skinner, thoughts and feelings, etc., were just other types of behavior to be explained.

- *Skinner did not believe in studying the brain to understand behavior.* Now, that's just silly. Skinner did not believe that the brain science of the twentieth century was advanced enough to provide much help to psychology. By the end of his life, he saw tremendous advances in neurobiology and genetics that suggested that that situation was changing.

- *Skinner opposed the use of theory.* Only partly true. Skinner opposed the use of theory that relied upon internal states like feelings and motivations. Operant theory explains behavior in terms of observable historical patterns.

- *Skinner opposed the use of statistics.* Actually, Skinner opposed the use of *group* statistics. Skinner felt that people should be studied individually so that our differences did not blot out our similarities.

- *Skinner thought all of psychology could be studied using rats and pigeons.* Almost. Skinner certainly felt that many human behaviors could be studied using nonhuman animals (including monkeys and apes). He felt that experiments with humans were less effective in identifying basic laws of behavior because of the influence of culture.

- *Skinner supported cures for mental disease using electric shocks.* Behavior therapy uses positive reinforcement.

- *Skinner wanted babies raised in operant chambers using electric shocks.* Now, really! Skinner actually opposed corporal punishment, including spanking, long before that view became popular. He did not oppose mild, brief punishment to stop severely autistic children from injuring themselves when nothing else worked.

continued ...

- *Skinner wanted the government to control everyone using conditioning.* Skinner felt that our behavior is controlled by our interactions with our environment. He argued that society would be better if people were less subject to random or unplanned events. Society as a whole, not the government, would establish systems for its own benefit using the principles of operant psychology.

There were even silly urban myths about B. F. Skinner, including one that he raised his second daughter in an operant chamber and that she spent her life in a mental institution as a result. Skinner did invent a special crib for his second daughter that included many gizmos to watch and play with, similar to the ones found in cribs today. She grew up happy and healthy and married and spent her life in London as an artist. She did see a counselor one semester in college (probably the same semester she had to break it to her parents that she was going to major in art).

COMPARING AND CONTRASTING CONDITIONING

There are many similarities and some important differences between classical and operant conditioning. Many of the same phenomena are found in both types of experiments. In fact, some have argued that there is only one type of animal learning (at least for nonhumans). They say that the two types of conditioning are just two different types of experimental procedure that generate learning in different ways. However, the differences between the two types of conditioning do make a big difference in terms of what types of behavior are learned.

Common Phenomena

There are a number of phenomena that can be produced using both classical and operant conditioning. Four of the most interesting are the following:

- **Extinction**   When the contingencies that conditioned the behavior—either the unconditioned stimulus in classical conditioning or the consequences in operant conditioning—are removed, the effects of the conditioning disappear, but only if the animal is presented repeatedly with the circumstances under which conditioning originally occurred. The animal learns that the bell no longer signals food or that pressing the lever no longer results in water.

- **Spontaneous recovery**   First, the response is completely extinguished. Then, the animal is removed from the setting where conditioning occurred. Then, after a time, the animal is returned to the setting and responding resumes.

- **Stimulus generalization**   The animal is conditioned in the presence of one neutral stimulus and responds (to a somewhat lesser degree) to similar stimuli. Behavior conditioned to a red light also appears in the presence of an orange light.

- **Stimulus discrimination**   The animal is conditioned in the presence of one neutral stimulus and is presented with similar stimuli without the contingencies that condition the behavior. Behavior conditioned to a red light is absent in the presence of an orange light when the orange light is presented alternatively during conditioning, but without the contingencies.

Many of the phenomena of conditioning can be thought of in terms of combinations of conditioning and extinction in various patterns. Similar stimuli share some elements, but not others. The lesser degree of responding in stimulus generalization may be due to a combination of conditioning to the shared elements and extinction for the unshared elements. In the case of stimulus discrimination, extinction to the similar stimulus extinguishes conditioning of all the shared elements as well.

Spontaneous recovery can be explained similarly. During extinction, the stimulus elements uniquely present soon after the animal is placed in the chamber may not be fully extinguished. When the animal is removed from the chamber and returned to it, those elements are present and the behavior spontaneously recovers.

Fixed reinforcement schedules produce pauses immediately after reinforcement. The stimulus elements uniquely present at that time are subject to extinction, because, in fixed schedules, a second reinforcement never immediately follows a first.

Differences Between Classical and Operant Conditioning

The first and most important difference between classical and operant conditioning is that classical conditioning conditions new stimuli to old responses, while operant conditioning conditions new responses to old stimuli. For instance, classical conditioning allows us to make salivation, eye blinks, and fear, etc., occur in the presence of arbitrary, new stimuli such as bells and lights and smells, but all of these responses are things the animal already did before we began the conditioning. In the case of operant conditioning, rats do not naturally press levers or pull chains until they are shaped to do so. In fact, it is sometimes difficult to operantly condition common, naturally occurring responses that are readily conditionable using classical conditioning techniques.

This points out a second difference between classical and operant conditioning. Classical conditioning works best with reflexive or involuntary responses. Operant conditioning, on the other hand, usually involves voluntary behaviors. In order to produce learning using classical conditioning, there must be an unconditioned reflex to work with. Operant conditioning can be used with any behavior the organism exhibits,

Tips on Terms: Response Generalization

Reinforcing one behavior increases the frequency of similar behaviors. This is called *response generalization*. Response generalization is not found in classical conditioning because the conditioned response resembles the unconditioned response so closely that genuinely novel responses do not appear. On the other hand, with operant conditioning, novel antecedent stimuli can be used for discrimination, which means that both novel stimuli and novel responses can be conditioned.

and novel behaviors can be generated using shaping because voluntary behavior tends to be more variable than involuntary behavior. Shaping relies upon the natural variability of all behavior.

A third difference is that classical conditioning is about the effects of the environmental stimuli that come *before* behavior, while operant conditioning is about the effects of the environmental stimuli (reinforcers) that come *after* behavior. This is also one of the biggest misunderstandings of Skinner's views on behavior (see "Variant View: Common Misconceptions About B. F. Skinner"). When people say that behaviorists believe that the environment causes behavior, we often misinterpret them to mean that the *immediate* environment (in the other words, the antecedent stimuli) causes behavior. Skinner believed that it was the long history of interactions between behavior and environmental *consequences* that shapes our behavior to make us who we are.

This leads us to the fourth difference between classical and operant conditioning. In classical conditioning, the role of the sensory stimuli that inform us about the world (the conditioned and unconditioned stimuli) is to trigger responses. In operant conditioning, the role of sensory stimuli (in other words, the discriminative stimuli) is to determine which response-reinforcer pairings we have experienced previously will affect our current behavior. For example, the stimuli found in yoga class signal that standing on our heads will bring praise. The stimuli found at the dinner table signal that standing on our heads will get us sent to bed without dessert.

Conditioned Reinforcers

Some conditioning phenomena can be explained as a combination of classical and operant conditioning. When the experimenter sets up the contingencies for classical conditioning, there is no way of preventing some operant conditioning from occurring, and vice versa.

One of the early arguments against Skinner's theory of operant conditioning was that not everything we do has an immediate consequence. How is it that we can perform such complex behaviors when the contingency isn't immediately experienced?

To counter this argument, operant theorists posited that there are *primary reinforcers* and *secondary reinforcers*. Primary reinforcers are things that directly satisfy a basic biological need. These would include food, water, and, as some have argued, sex. A secondary reinforcer is a stimulus that leads to a primary reinforcer. With conditioning, the secondary reinforcer can acquire reinforcing properties. Alone, it does not have any intrinsic value to satisfy a biological need.

Operant theorists argued that people maintain long chains of complex behaviors by chaining together secondary reinforcers. These serve as placeholders for long sequences of behavior that eventually result in a primary reinforcer. For example, although people derive many kinds of satisfaction from their jobs, the behavior of working is maintained, in part, by getting a paycheck on payday. The behavior of going to the bank and cashing the check is maintained by obtaining paper money. The behavior of spending that money in the grocery store is maintained by taking home food. The behavior of cooking a meal is maintained by eating. Since eating is the only primary reinforcer in this long sequence of behaviors, all of the other behaviors are maintained by things that have no direct reinforcement value to satisfy a biological need. Thus, the secondary reinforcers maintain the chain of behavior until a biological need can be satisfied.

Note that secondary reinforcers begin as neutral stimuli and then acquire the reinforcing properties of primary reinforcers. Note also that primary reinforcers, such as food, tend to be stimuli that have unconditioned responses, like salivation, associated with them. Some theorists argue that the regular pairing of secondary reinforcers with primary reinforcers during the operant chaining produces classical conditioning of the secondary reinforcers. In other words, secondary reinforcers are reinforcing because they signal the availability of primary reinforcers via classical conditioning.

The distinction between primary and secondary reinforcers is not always clear. When we check out the news (behavior), we are informed about the day's events (consequence). Being informed is a secondary reinforcer because it enables us to make better decisions and to be more entertaining in our dinner table conversations. But being informed has been useful to animals for millions of years. It may be a primary reinforcer. The love of knowledge may be as much a part of our genes as the love of food.

Cognitive Learning

The 1960s brought the cognitive revolution in Western psychology and, with it, an alternative theoretical view on learning. Many contemporary psychologists rejected the traditional behaviorist perspective because it did not take the organism's thoughts and perceptions into account. This is not to say that the cognitivists argued that learning did not occur in an operant or classically conditioned manner. They thought that it was incomplete in that it could not account for all learning processes involved. The cognitive view of learning holds that in almost any associative learning situation, mental processes, or cognitions, intervene between the stimulus and response. Cognitive theorists also argued that some types of learning could not be accounted for solely by classical and operant conditioning.

DEFINITION

In order to explain the role of these other processes in learning, psychologists have constructed a precise definition of learning. The definition of learning has four components. The first component is the concept of *performance*. Early behaviorists argued that the way to study learning is to infer it from changes in performance. They create a controlled situation that is conducive to learning and then objectively measure performance at different times. If performance changes, then they are able to claim that the change indicates learning.

Defining learning by performance change alone is problematic because it does not account for other factors, such as motivation or emotion. It also does not account for the fact that even if something is learned, a performance change may not occur right away. A definition of learning must also account for the *potential* for performance change.

The third component involves the *permanence* of the potential performance change. This is done to distinguish learning from other factors, such as emotion or maturation, which are relatively fleeting. Finally, the definition of learning includes the concept of *experience*. This distinguishes learning from physical factors that influence performance change, such as maturation or illness. Learning results from experience. Given all of these elements, contemporary psychologists define learning as a relatively permanent change in performance potential that arises from experience.

EXPECTATION

One of the ways of thinking about learning potential is the notion of *expectation*. Tolman, who considered himself a behaviorist, was one of the first to challenge strict behaviorism on this issue. He argued that many animals are capable of thinking about the consequences of their behavior and choosing the most rewarding action in a purposeful way. Central to this view is that the animal can form a mental representation of how things in the environment typically respond and how they are related to each other. It is this expectation that brings about the response.

> ## Case Study: If Music be the Love of Food...
> Tolman used an anecdote from Pavlov's own lab to demonstrate his argument. A dog had been classically conditioned to salivate to a ticking metronome by the association of the ticking and the food. When the metronome was turned off and the dog released from the apparatus, it sat in front of the metronome and whined and begged. The dog *expected* the metronome to provide more food.

LEARNING WITHOUT ASSOCIATION

Over the decades, cognitive psychologists have provided several lines of research in which a learned response was neither classically nor operantly conditioned. Evidence that learning can occur without reinforcement supports the cognitive perspective of learning. We discuss two major examples here.

Latent Learning

Latent learning occurs when an organism learns a new behavior but does not perform that behavior until there is an incentive. For example, a person can learn the way to an unfamiliar part of town if someone else simply tells her how to get there. The route is stored in memory as a sort of cognitive map and is used when the need arises. Reinforcement and/or practice are not necessary for latent learning to occur.

Social Learning Theory

Social learning theorists argue that humans acquire a large variety of strategies, perspectives, and rules about behavior that we imitate, avoid, or modify to our advantage. They further argue that we do this in the natural course of cognitively processing what goes on around us and thus do not require reinforcement. For example, Bandura and associates demonstrated that people learn a variety of behaviors by *modeling* the behavior of others. In a classic series of experiments, they allowed children to watch an adult play with a toy. The adults played with the toy in a manner that was either aggressive or gentle. When the kids were allowed to play with the toy, they imitated the adult's behavior. Thus, the children learned how to behave with the toy by watching someone else.

Reward and punishment do have a place in social learning theory, but they hold a broader place than in strict behaviorism. Social learning theorists' main claim is that

Fun Facts: Modeling Behavior

Modeling behavior is more prevalent in children than in adults, but not for the reasons you might think. Some mistakenly think that modeling is something we grow out of as we mature. However, adults are just as likely to model the behavior around them as are children. The important factor to consider is *when* a person, child or adult, needs to model behavior. The key is that we are likely to model behavior when we are in an unfamiliar situation. As children, so many situations are unfamiliar because our world is expanding. As adults, we spend most of our time in familiar settings, like work, home, school, etc. However, when you expose an adult to an unfamiliar situation, they will be just as likely to model the behavior around them. Think about the first time you were traveled to another country, sat in a college classroom, or went to a party where you did not know a lot of people. What did you do?

reward and punishment are important to learning, but they are not *necessary* for learning to occur. For example, a child watching her older brother clear the dinner dishes will add this behavior to her repertoire of possible behaviors. If she also sees that her parents praised her brother for this behavior, she will be more likely to enact this behavior in the future. Thus, reinforcement will affect the likelihood of performing a behavior, but not necessarily the learning of the behavior itself.

Vicarious and *intrinsic* reinforcement and punishment illustrate this point. In the previous example, the little girl learns the positive consequences of clearing the dinner dishes merely by watching her older brother. To learn vicariously is to be influenced by the consequences we observe others receiving for their behavior. If we see someone being rewarded for his behavior, we are more likely to enact that behavior. If we see someone being punished for his behavior, we are less likely to enact that behavior. However, the consequences in the environment are not the only reinforcers and punishers that affect the likelihood of enacting a behavior.

Tips on Terms: Reward

Note that cognitive learning theorists, unlike behaviorists, use the term, *reward*. They find the term acceptable because they believe that the intrinsically rewarding properties of the reinforcing stimulus are key to ensuring that learning occurs. Consequences are reinforcing or punishing because they produce pleasant or unpleasant feelings and those feelings motivate the person. Strict behaviorists doubt the necessity of motivations or emotions in making learning happen.

Our own internal, or intrinsic, standards of behavior affect this as well. We each have internal standards of how we want to perform on an exam, for example. If we feel that we met or exceeded these standards, we feel good about ourselves. High self-esteem is a powerful reward and will increase the likelihood that we will enact the behaviors that lead us to perform well. If we feel that we failed to meet our standards, we feel bad about ourselves. Low self-esteem is also a powerful punishment and will decrease the likelihood that we will enact the behaviors that lead us to perform poorly. Thus, our intrinsic standards of behavior serve as both reward and punishment, even before we receive our grade.

Ethology

Ethologists feel that both cognitivists and behaviorists neglect the genetic components of behavior and learning. They have shown two important features of learning due to genetics. First, they have shown how genetically preprogrammed behaviors, called *fixed action patterns*, serve as the basic building blocks of learning. Second, they have demonstrated important differences between the ways that animals of different species learn.

How is it that different types of birds make different types of nests and sing different birdsongs? Is the entire procedure for building a nest encoded in their genes? Do the mother and father birds teach the baby birds to sing? As it turns out, these very complex behaviors are intricate combinations of learned and innate behaviors.

Do squirrels memorize the location of every single nut they bury in the fall so they can eat through the winter? No. They have specific innate behaviors that make some types of places more likely places to bury nuts. They return to the most likely places often enough to survive the winter.

Ethologists have investigated how bees learn the way from the hive to the flowers and how they teach the other bees the way by dancing for them. The first bee learns from experience. The other bees learn from the first bee's dance. They have also shown how different types of animals are preprogrammed to learn in different ways. Many animals return to places where they have previously found food. This strategy, called *win-stay*, works well for predators (because prey animals tend to revisit the same places) and can be explained in terms of reinforcement. Other animals avoid places where they have previously found food. This strategy, called *win-shift*, works well for foragers (because once you have dug up all the nuts or eaten all the berries in one spot it is a while before they are replenished) and is harder to explain in terms of reinforcement.

Computational Approaches

Scientists often use computers to show how their theories will work out. The computer software application programs, called *models*, simulate the effects of the theory the way computer games simulate virtual worlds. If the virtual world matches the real world, that is evidence in favor of the theory.

In psychology, there are two basic types of computer models: *symbolic* and *connectionist*. Symbolic models began at the start of the cognitive revolution. Like cognitivism, they had great difficulty modeling learning early on. More recently, symbolic learning has made important advances. Connectionist models became popular only in the 1980s. They combine cognition and learning in one model. Because connectionist models are designed to resemble groups of interconnected nerve cells, they are often called *neural networks*.

Summary

As we have seen, the study of learning involves a number of different approaches. Ethology tends to address different issues with a different focus. The computational approach offers tools that can be used with any other approach. As such, both of these approaches can be seen as complementary, rather than competitive with the other approaches. However, the behavioral and cognitive approaches offer competing explanations of much of the same phenomena. The strengths and weaknesses of these two competing approaches are worth looking at.

The most important difference between the two approaches is that learning itself plays a different role within psychology in each. For the behaviorist, learning is central to psychology. It is assumed that basic principles of learning underlie many other psychological phenomena, including thinking, social psychology, language, personality, psychological disorders, and so on. Because none of these behaviors is present at birth, it is assumed that they are learned. For the cognitivist, it is thinking, understood in terms of information processing, that is central. We are all assumed to share a basic mechanism for coping with the world. Social psychology, language, personality, and learning, are all just applications of the mechanism that underlies thinking.

From the cognitivist perspective, the biggest weakness in behaviorism is that it leaves out much of what common sense tells us is involved in learning. Motivated students learn more. Positive and negative consequences feel different and give rise to different emotions. Memorizing facts for a test in school doesn't seem to work like conditioning and doesn't seem to involve consequences or the emotions that go with them. (Of course, this last observation is being made by folks who all finished high school, have never been at risk of flunking, and learned very early in life to avoid the punishment associated with not doing one's homework.) Cognitive learning theory tries to integrate these other psychological elements into the study of learning.

From the behaviorist perspective, the biggest weakness in cognitivism is that it appears to have some logical circularity. If learning is explained by storing and retrieving and manipulating representations of things in the world, how are representations learned? If we learn that elephants are gray by associating the concept of elephant with the concept of gray, where do the concepts of elephant and gray come from? If consequences are effective motivators because we understand them beforehand, how do we learn about consequences? Don't we have to learn about something in order to understand it?

In short, the role of learning within the rest of psychology is a key difference between the competing schools of thought in contemporary psychology. A fully scientific psychology will be able to explain learning within the context of the study of psychology without these conceptual problems.

Quiz

1. Which of the following is/are an assumption of behaviorism?

   (a) Behavior is determined in some way.

   (b) Explanations of behavior based on internal causes are useless.

   (c) The environment shapes behavior.

   (d) All of the above.

2. In basic classical conditioning experiments, the bell is the _____.

   (a) Unconditioned stimulus

   (b) Unconditioned response

   (c) Conditioned stimulus

   (d) Conditioned response

3. _____ conditioning is the most efficient way to establish a classically conditioned response.

   (a) Forward

   (b) Backward

(c) Shaping

(d) Random

4. In operant conditioning, spanking a child when he talks back to his parents is a form of _____.

   (a) Negative punishment

   (b) Positive punishment

   (c) Negative reinforcement

   (d) Positive reinforcement

5. Reinforcement given after the first bar press after an average period of time has elapsed within the experimental session is a _____ schedule.

   (a) Variable-interval

   (b) Fixed-interval

   (c) Variable-ratio

   (d) Fixed-ratio

6. _____ occurs when the contingencies that condition the behavior are removed and the effects of the conditioning disappear.

   (a) Stimulus discrimination

   (b) Stimulus generalization

   (c) Spontaneous recovery

   (d) Extinction

7. _____ conditioning involves reflexive behaviors and _____ conditioning involves voluntary behaviors.

   (a) Classical, classical

   (b) Operant, classical

   (c) Classical, operant

   (d) Operant, operant

8. Which of the following is not a component of the definition of learning?

   (a) Performance

   (b) Permanence

   (c) Experience

   (d) None of the above

9. According to _____ theory, people learn a variety of behaviors by modeling the behavior of others.

   (a) Classical conditioning

   (b) Social learning

   (c) Operant conditioning

   (d) Ethology

10. Ethologists argue that the win-stay strategy works well for _____, and the win-shift strategy works well for _____.

    (a) Foragers, foragers

    (b) Foragers, predators

    (c) Predators, foragers

    (d) Predators, predators

CHAPTER 12

Psycholinguistics: The Psychology of Language

Almost all animals communicate. Many communicate using sound. The principal mode of human communication has an apparently unique combination of special characteristics and so deserves a special name and separate study. We call this mode of human communication *language*.

Overview

The value of communication to animals is obvious. When one animal detects danger, other animals can escape without having to detect it themselves. All they have to do is hear the warning cry. One animal can communicate the availability of food or water or of themselves for sex to other members of the group, and so on. With the enormous complexity of the human mind comes an enormously powerful means of communication called language. It not only serves the ordinary purposes of animal communication extraordinarily well, it also serves a huge host of other purposes as well.

Oral communication appears to have evolved as a part of human behavior, or at least, if there was an inventor, she predates human history by far too long to be known by us. Although we do not know the inventor of written language, it appears to have been invented by someone. Written language gives permanence and portability to the same sorts of messages we can communicate orally. The hearer, now a reader, can be far from the speaker, a writer, both in time and space.

HISTORY

The history of the study of language is very old, although not as old as language itself. After all, the study of anything more or less requires language, which is one of the more remarkable things about language. Language allows us to think about, and pass along our knowledge of, almost anything, including language itself.

Concern with language dates back at least to early Greece and India. Much of this study concerned language itself, its grammar, its logic, how it should be written or spoken, and so on. Psychologists are more concerned with how people *use* language than with either the rules or form of language itself. However, in order to study language, *psycholinguists* (the name given to psychologists studying language) have to understand a great deal about what philosophers, logicians, linguists, grammarians, sociolinguists, and even literary critics have discovered about language.

The scientific study of language, both linguistics, in which language is considered on its own, and psycholinguistics, in which language is considered in terms of behavior, began about a hundred years ago. Contemporary experimental psychology found very little of use in the linguistics of the time. In addition, the complexity of language discouraged early experimentalists, who found plenty of simpler, understudied phenomena to work on. An important exception was the study of the child's development of language. As we will see, the dramatic changes from the nonverbal infant to the verbal adult provide a wealth of phenomena to study now, just as they did then.

As the twentieth century progressed, experimental psychology was moving in the direction of behaviorism, and linguists understood language to be about the expression of ideas, a type of internal state that behaviorists were uncomfortable with. Attempts to study language without the notion of ideas were limited during the behaviorist age. Behaviorally inclined psychologists attempted to understand language from the bottom up. They did memory studies with nonsense syllables. They investigated word frequencies in various texts, such as novels and technical manuals. But these studies focused on the form of language, rather than the meaning conveyed by linguistic communication. It is very difficult to work with meaning while avoiding the notion of ideas.

CURRENT APPROACHES

The cognitive revolution of the 1950s relegitimized the notion of ideas within psychology. While psychologists began to find ways to study thinking experimentally, Noam Chomsky, a linguist, developed the concept of a *representation*, a hypothetical brain state that acts like a symbol, standing in for something in the real world. Presumably, when we are thinking about dogs, or cats, or tables, or Bizet's *Carmen*, part of the brain is activated in a pattern corresponding to the appropriate symbol. In this way, different brain states have different meanings the same way that words do.

Just as psychologists were profoundly influenced by Chomsky's ideas about thinking, the psychologists who studied language were profoundly influenced by Chomsky's linguistics. Much of current psycholinguistics studies how people communicate with language structured according to Chomsky's theories. Almost all of it assumes that language and thought are organized according to these principles.

Also in the 1950s, quite independently of the cognitive revolution, one of the most prominent behaviorists, B. F. Skinner (discussed in Chapter 11) attempted to explain meaning in language in behaviorist terms. He attributed the meaningfulness of language to operant conditioning with various components of language acting as both responses and stimuli. Language, which Skinner referred to as verbal behavior, has meaning because it is used by people to obtain reinforcement via the behavior of other people. When we need salt and cannot reach it ourselves, we ask someone at the other end of the table to pass the salt, and she does. If playing with the cat will be positively reinforcing, but we haven't seen the cat recently, someone else telling us that the cat is on the mat enables us to obtain this reinforcement. Skinner's view of language was that it is a means evolved so that groups of people can coordinate their behavior to accomplish what one person cannot do alone.

Currently, the cognitive view of language dominates research. Few researchers follow Skinner's behavior analytic approach. In addition, the study of language is one of the many areas where new methods of observing brain activity are being used, mostly to identify what parts of the brain are used in different aspects of linguistic communication.

The Nature of Language

According to the cognitivist view, language is a collection of symbols that convey meaning, plus the rules for combining these symbols, that can be used to generate an infinite variety of messages. Not all forms of communication are considered to be language. Whether written or spoken, a communication system must meet four criteria in order to be classified as a language:

- **Symbolic** Elements of the system enable reference to aspects of the world that are distant in time or place or even only possible.

- **Semantic** Which elements of the system relate to which aspects of the world (or have whatever other sorts of meaning) is set by social convention and is unrelated to the form of the elements.

- **Generative** Elements of the system can be combined according to rules to generate different meanings.

- **Structured** The order in which the elements are presented matters.

First, language is *symbolic*. Words, whether spoken or written, are used to symbolize or represent actions, objects, events, and ideas. They allow us to refer to objects that are not present right now, to things that happened in the past, or that may happen in the future. Symbols expand what we can communicate about. For example, the word *chair* refers to a class of objects that have certain properties. If I say this word, you understand that I'm talking about something you can sit on. You may not know the details of the chair, but you understand the general gist.

Second, language is *semantic*, or meaningful. There is no built-in relationship between a word and what it symbolizes. So, a language can have different words to represent a concept. Because words are arbitrary, they need to have a shared meaning for the speakers and listeners of the language. For example, all speakers of the English language must agree that the word *chair* represents a certain kind of object used for sitting and that *on* represents the position of the person with respect to the chair when the chair is in use. The spelling and sound of *chair* and *on* don't tell us the meaning.

Third, language is *generative*. Symbols can be combined in an infinite variety of ways to generate new messages. We use the symbols to create and understand sentences we've never spoken or read before. For example, before opening this book, you had probably never spoken or read the sentence, "I did not come all the way to the Moroccan desert so I could sit there eating couscous like the other rubes." Yet, you understand this novel sentence. You can parse its meaning even though you have never seen the sentence before.

Finally, language is *structured*. There are rules that govern how we string words together. Some arrangements of words are acceptable and some are not. We rely on this structure to understand the meaning of the sentence or phrase. For example, take the sentence, "The kitchen into the rushed cook." Here, we read this as if the *kitchen* is taking the action, the *cook* is being acted upon, and *rushed* is an adjective describing the cook. Our reading is based on the grammatical rules used to structure the English language. If we used the correct structure, however, we would reorder the sentence to read, "The cook rushed into the kitchen." Now, it makes sense.

The structure of language corresponds to grammar. There are rules for what is correct in ordering sentences. Psychologically, a sentence may have bad grammar and still be understandable. If a sentence has good grammar, but doesn't make sense, then something else is wrong. It may be the semantics, as in the grammatically correct sentence, "Green ideas sleep furiously." Or it may be in what is called the *pragmatics*. Pragmatics involves the social rules for communicating. For example, there is nothing wrong with the sentence, "My child is an orphan." However, there is something wrong with *saying*, "My child is an orphan."

Sometimes, the actual meaning of a statement is completely changed by the circumstances under which it is spoken. Imagine two neighbors sitting in a kitchen on a warm, sunny day. The first neighbor's child walks in from outdoors, leaving the door open, and heads for the living room. The parent says, "Gee, isn't it cold in here?" Translation: "Turn around, march yourself back to the door, and shut it!"

Variant View: Do Animals Have Language?

Various nonhuman animals exhibit communicative behavior that resembles language. Birds sing, bees dance, dogs respond to commands, parrots mimic human speech, and so on. Some of these animals can be trained to understand and communicate in ways that capture some, if not all, of the criteria listed for language.

Bonobos (formerly called pygmy chimpanzees) display a remarkable ability to deal with semantics. Lacking a vocal apparatus capable of the range of sounds necessary to human language, bonobos are trained to speak and respond using boards with symbols. By pointing at a symbol on the board, the bonobo or the trainer can request or refer to things not present in the immediate environment.

Dolphins have been trained to communicate with humans in a way that respects grammatical order. They can distinguish between commands to bring the ball to the hoop and to bring the hoop to the ball.

Possibly the most fascinating work with animal language is Irene Pepperberg's work with African Grey Parrots. Beginning with an African Grey named Alex, Pepperberg developed a complex systematic series of training procedures to establish whether or not parrots could communicate in ways that demonstrated all four of the cognitivist criteria for language. The question of whether all four criteria have been met is still controversial, but Pepperberg has clearly demonstrated that parrots are capable of much more than mimicry and are very close to being able to communicate in ways that meet all four criteria.

Language Development

Although the human capacity for language begins to develop very soon after birth, infants come into the world completely speechless. In a matter of a few years, all children of normal development are highly skilled speakers and listeners. How do they progress in such a short time? Although children grow up in vastly different cultures with vastly different languages, they seem to go through very similar sequences in learning to speak their native language. This may reflect the maturation of the areas of the brain that process language.

From birth to one year of age, infants are preverbal, meaning that they are not yet able to communicate using words. This does not mean that they do not communicate. In the first three months, newborns communicate through cries that vary in tone and rhythm, depending on how the infant is feeling. The cry for food is different than the cry for a clean diaper. Most parents learn to distinguish these cries so they can tend appropriately to the infant's needs.

During the next three months, an infant begins to coo both alone and in "conversations" with adults. If an adult says something to the infant, the infant will coo in response and then pause as if waiting for a response. This turn-taking pattern has the structure of a real conversation. Infants who experience this pattern of turn-taking tend to sound very different from babies who do not. The cooing sounds more like real syllables. Thus, they may be picking up the structure of linguistic conversation before they are able to understand meaning.

In the last six months of the first year, infants begin babbling. They are able to chant various syllabic sounds in a rhythmic fashion. Older infants also communicate through a combination of gross motor gestures and voice intonation. Thus, they are better able to copy adult speech patterns as they get older.

By the end of the first year, children know the names of a few people and objects and begin to formulate their first words. To reach this stage, they must understand that sound can be used to convey meaning. Simply saying the word in the right context is not enough. The child must understand the meaning. Their first true words usually refer to immediately tangible and visual objects or people. They can label people, objects, and actions present in the here and now.

A child's vocabulary starts expanding rapidly at about two years of age. However, even though they may not use all of the words they know, they can understand a lot of words. Their *receptive vocabulary*, or the words they understand, is greater than their *productive vocabulary*, or the words they produce. The key to the rapid growth in vocabulary between one to six years of age involves *fast mapping*. This is the process that children use to map a word onto the underlying concept after only a single exposure to the word. Children are very adept at figuring out what concept a word symbolizes.

However, children are not always correct in their word usage. They make errors. But even these errors follow certain patterns. Children have a tendency to overextend the meaning of words. They will use a word for a broader class of objects than what the word symbolizes. For example, a child who has learned that *dog* refers to the family Dalmatian may apply that word to every animal he sees. Conversely, children also have a tendency to underextend the meaning of words. They will use a word to mean only the original object and not other similar objects. For example, the child may use *dog* to refer to the family Dalmatian but not the first poodle she sees. Through over- and underextension, along with feedback about mistakes, a child gradually develops the same mental representation of meaning that others have.

Once children have acquired a basic vocabulary, they begin to use *holophrases* to convey meaning. Holophrases are single words that are used to convey a phrase or sentence. For example, a child may say "Book" to mean "Read me the book." Some theorists think that this is the transitional step between a child's first words and the use of their first sentences. The formation of first sentences must wait for a certain level of neurological maturation, including a rapid increase in the number of synapses in the cerebral cortex.

First sentences are typically very simple two-word phrases limited to concrete nouns and action verbs. These are referred to as *telegraphic speech* because the terseness of the speech sounds like the wording of a telegram. Even though first sentences are relatively simple, they are structured. The words are combined in an order that follows adult syntax. Thus, a child will say "eat cookie" instead of "cookie eat." The impressive thing about telegraphic speech is that a child can convey a wide range of meaning using simple two-word sentences. See Table 12-1 for examples.

Table 12-1   Range of Meaning for Telegraphic Speech

Sample Sentence	Meaning
Mommy hug.	Agent-action
Kitty furry.	Attributes
That Daddy.	Identification
Kitty here.	Location
No bath.	Negation
All-gone milk.	Nonexistence
My blanket.	Possession
Where kitty?	Question
Tickle again.	Recurrence

By the time children enter school, most have a good grasp of the grammar of their native language. They acquire an implicit sense of how to organize words into increasingly complex sentences. Also, grammatical rules are learned in a fairly predictable order. One process is through the use of over-regularizing grammatical rules to instances where they do not apply. This is how children learn forms of the English past tense. They first use certain irregular past-tense verbs (for example, *came*, *fell*) correctly. When they learn the rule for forming regular past tense verbs (add a *d* or *t* sound), they try to apply it to irregular verbs as well (for example, *goed* or *falled*).

Something Extra: Bilingual Children

Some children are raised learning more than one language. Does this give them any advantage or disadvantage? What about adults who learn a second language? What does the study of multilingual people tell us about language development?

Bilingual children seem to do better on some cognitive tests. However, it is hard to say whether this is a result of their bilingual upbringing (which may have provided more cognitive challenges) or whether it is just that the least cognitively able children are not raised bilingually. At any given age, a bilingual child's total vocabulary is about the same as that of other children, but this means that she knows fewer words in any one language than native speaking children of that same age.

Of course, as those of us who have studied a foreign language know, adolescents and adults do not just acquire additional languages the way bilingual children do. The fact that learning a language as an adult requires study leads to the notion of a *critical period*, a specific age range when language, or some aspects of language, is learnable. Studies have been done with people of various ages who have immigrated to a country where they had to learn a new language. Their final competence, particularly their command of grammar, is related to how old they were when they immigrated. The younger they were, the better their acquisition of the second language. After the age of nineteen, however, things level off. Older adults learn new languages just as well as younger adults.

Even though there is a special time for learning language, there appears to be no special place. In the normal brain, there are specific parts that are used for language. If a child has damage to those parts of the brain before he learns language, other parts of the brain take over, and there may be almost no language deficits. If an adult has damage to the parts of the brain that control language, it may be impossible for her to relearn her own language, or at least some aspects of it.

Tips on Terms: Child-directed Speech (CDS)

In many cultures, adults speak very differently to young children than they do to one another. They use short sentences, simplified vocabularies, childlike mispronunciations, refer only to objects that are present, and use a high-pitched voice with a sing-song rhythm. It has been argued that this *child-directed speech*, formerly called "motherese," helps the child learn language, in part because the sing-song sound maintains the child's attention and in part because the simplified nature of the speech provides the child with a stepping stone to more complex speech.

Others argue that CDS usually has complex grammar, which is the hardest part of language to learn. They also note that many cultures do not use CDS. Adults speak to children the same way they speak to other adults. There is no good evidence that the absence of CDS slows or limits the child's acquisition of language.

During this time, children also start to learn the pragmatics of language. This involves using language appropriately within a given social context. Children learn to shape their use of language to suit the social situation. For example, when a six-year-old speaks to a two-year-old, he will use a lot of attention-getting words that he wouldn't use when speaking with an adult. The six-year-old realizes that different listeners require different kinds of speech. The child will also begin to pick up on the rules of conversational language, such as turn-taking.

Language Acquisition

The theories of language acquisition are deeply rooted in the nature-nurture question. Some theorists argue that language acquisition is largely genetic, others argue that environmental factors are largely at work, and still others argue that it is a combination of both.

The nativist theories argue that language acquisition is controlled by the genetically programmed development of certain neural circuits in the brain. Chomsky argues that the brain has a structure that automatically analyzes the components of the speech a child hears. This *language acquisition device* is believed to be present in all humans, but not in other species.

There are several lines of evidence that support the nativist theory. First, children acquire very different native languages in much the same way. The processes of language acquisition and development for American English and Cantonese appear

Critical Caution

The rapid and effective acquisition of language by children without explicit teaching applies only to *spoken* language. Reading and writing must be taught. If there is a language acquisition device that allows us to learn our first language without explicit teaching, the nativist theory must also explain why it doesn't work for written language.

to be the same. Thus, some conclude that humans may be born with innate tendencies to structure language in specific ways. Second, children are raised in a wide variety of learning environments, yet they acquire the same level of linguistic skills at about the same age. Finally, children acquire a vocabulary of approximately 14,000 words between the ages of two to five years. This rapid rate of language acquisition would seem unlikely without some innate propensity to master speech.

The learning theorists argue that no child will ever naturally acquire a language without being exposed to the language first. Thus, some learning must be involved. Some parents believe that praising children for correct grammar and expressing disapproval for mistakes will help to improve their children's language. Studies indicate that it does have an effect, but not to the extent that you may think. In reality, parents naturally correct their children for meaning rather than for grammar. They are more concerned with the correctness of the meaning children are conveying rather than the correctness of their grammatical statements. Reinforcement learning has some effect, but it appears that it cannot completely account for language acquisition.

Learning theorists have also examined whether modeling can account for language acquisition. Do children merely imitate sentences they hear others speak? Anyone who has let an "adult" word slip out in front of a child can tell you that children do imitate what they hear. However, the evidence suggests that they do not rely on imitation solely. Children are able to produce sentences that are novel. They can take the words they know and, using the meaning they convey, put them together to build sentences they have never heard before.

The interactionists or integrative theorists argue that biology and environment interact to contribute to language acquisition. Language development is tied to cognitive development, which also depends on both maturation and experience. For example, children are able to learn the rules for past tense only when they are capable of understanding the concept of the past. Language development is also facilitated by the interaction with more experienced speakers. Children use the feedback they receive to improve their use of pragmatics.

Variant View: Linguistic Relativity

Benjamin Whorf advocated an extreme version of the nativist thesis, called *linguistic relativity*. He argued that the language we learn first actually shapes the way we think about and even perceive the world. In cultures where there are many more words for, for example, *snow* or *rice*, than there are in English, do the native speakers acquire an improved ability to perceive and understand subtle differences in snow or rice?

Overall, the evidence has not supported Whorf's thesis. The Dani of New Guinea have very different color names than we do, but their perception of color is very similar to ours. There is one intriguing exception, of a sort. The language sounds, or phonemes, of different languages are organized differently. In Spanish, the sound for *s* includes both the English sounds for *s* and *z*. Not only do native Spanish speakers tend to use a sound between *s* and *z* when pronouncing either sound in English, but adult native Spanish speakers actually have trouble hearing the difference between *s* and *z* even when spoken by native English speakers.

This is not a matter of the complexity of the language. If you are a native English speaker, say "Block that kick!" out loud. How many different *k* sounds do you hear? Actually, there are three. Just as native Spanish speakers cannot hear the difference between *s* and *z*, which you can hear easily, you cannot hear the difference between those three *k* sounds. Speakers of some other languages, including some Arabic languages, can hear the difference easily, because their native languages have all these different phonemes, just as English has both *s* and *z*.

The structure of the language we learn as children may not affect the way we perceive the world, but it does affect the way we perceive the sounds of language.

Other Forms of Communication

Humans communicate using methods that don't seem to match the criteria for being a language. What are these other means, and are they languages?

People use *nonverbal expressions, gestures*, and *body language* to communicate. The distinctions between these are not hard and fast. Nonverbal expressions, such as facial expressions and tone of voice, etc., can occur either with or without verbal communication. When they occur together with verbal communication, they are used to enhance what is being said, adding emphasis, indicating emotional content such as doubt or surprise, or even irony, suggesting that what is being said is not

Extra Exercise

Try the following on a friend. On a piece of paper, write down the words, "You will use your hands." Fold up the paper and put it in front of your friend. Ask your friend to define either a spiral staircase or a goatee (a.k.a. a Van Dyke). Unless they know the trick, they will almost certainly use their hands when describing either one. Then have them open up the piece of paper and see how clever you are.

literally what is meant. On their own, they are used to substitute for verbal communication. A parent may use a stern expression without saying, "Stop that!" in order to indicate, "Stop that!" Gratitude or sympathy, etc., may be indicated verbally or nonverbally or both.

Gestures always accompany language. They not only enhance what is being said, but also clarify. Intriguingly, gestures may have a genetic component. Persons blind from birth tend to use the same hand gestures as sighted persons. They also use them even when speaking to another blind person.

Body language is a term coined to describe nonverbal communication that conveys things that the speaker does not necessarily want to convey. Whether accompanying verbal communication or not, the body language message is not part of the verbal message. In general, body language tends to convey attitudes toward the situation or the others present, or general personality characteristics. Friendly speech can be belied by nervous or hostile body language. Persons highly skilled in nonverbal communication can and do use their ability to interpret body language to their benefit. Good salespersons can spot potential prospects. Muggers can pick out people less likely to put up a fight. Effective mating behavior usually involves both parties being able to detect the other person's sexual interest, in part with body language.

In none of these cases does the communication match all four criteria of being a language. Theoretical criteria aside, these modes of communication either are used to enhance linguistic communication or have a very limited variety of things they can express. Someone's body language may be able to tell you she wants you to leave, but not whom she would rather see. At a buffet, it might tell you someone prefers the sausages to the salad if both are in front of him, but it tells you nothing about how he might feel about butterscotch pudding if it is not at the buffet table.

This doesn't mean that nonverbal communication is unimportant. All of our experiences on the telephone show us how much harder it is to communicate without using gestures or facial expressions. And our experiences with email (or book-writing)

Big Background: Language and Brain Regions

Based on studies of brain-damaged patients, it has long been known that most people use one side of their brains, usually the left side, for some critical language functions. Now that we have a number of new ways of scanning the brain in detail while people are behaving, there has been an explosion of new research attempting to discover just where in the brain the activities underlying various aspects of language happen.

The theory is that, because language is such a highly specialized function, specific areas of the brain have evolved to handle many of the specific functions of language. Researchers in *cognitive neuroscience* give people various tasks to perform while the researchers scan their brains. By carefully selecting similar tasks that differ only by one linguistic function, researchers hope to identify a brain region that turns on and off when that function is and is not used. While initial results are promising, this is a very new area of research using very complex techniques to study very complex phenomena.

show us that it is even harder to communicate when we can't use various tones of voice.

Finally, we should consider sign language. Some hearing-impaired persons use a gestural language, as opposed to an auditory language, to speak. The best known of these is American Sign Language, or ASL. Unlike nonverbal expressions, gestures, and body language, sign languages are true languages with grammar, semantics, and all the other features of true languages.

Summary

Language is an enormously complex phenomenon. In some ways, it is almost as complex as all of the rest of behavior combined. In addition, many other areas of psychology involve language and even more require language for their study. Psychologists have worked hard to find empirical ways of investigating many different aspects of language. Because language is a field of study in its own right, and has an important role in other disciplines outside of psychology, many of the theories tested by psychological methods actually come from linguistics and other areas outside of psychology. In addition to providing empirical evaluation of theories of language, psychologists study language as a form of behavior, a form of behavior used for communication.

Quiz

1. Which of the following is a component of a communication system, or language?

    (a) Symbolic

    (b) Generative

    (c) Structured

    (d) All of the above

2. A _____ vocabulary is the words we understand, and a _____ vocabulary is the words we generate.

    (a) Receptive, productive

    (b) Productive, receptive

    (c) Receptive, receptive

    (d) Productive, productive

3. Children use _____-extension to gradually develop the same mental representation of word meaning that others have.

    (a) Over

    (b) Under

    (c) Both a & b

    (d) None of the above

4. _____ is/are single words that are used to convey a phrase or sentence.

    (a) Telegraphic speech

    (b) Over-regularization

    (c) Cooing

    (d) Holophrases

5. The _____ of language involves using language appropriately within a given social context.

    (a) Pragmatics

    (b) Semantics

    (c) Grammar

    (d) Meaning

6. The _____ theorists argue that language acquisition is controlled by the genetic programmed development of certain neural circuits in the brain.

   (a) Learning

   (b) Nativist

   (c) Cognitive

   (d) All of the above

7. The _____ theorists argue that biology and environment interact to contribute to language acquisition.

   (a) Nativist

   (b) Learning

   (c) Integrative

   (d) Social learning

8. Which of the following is true about the use of gestures?

   (a) Gestures can accompany verbal communication.

   (b) Gestures can be used without verbal communication.

   (c) Gestures may have a genetic component.

   (d) All of the above.

9. _____ is a term coined to describe nonverbal communication that conveys things that the speaker does not necessarily want to convey.

   (a) Gestures

   (b) Body language

   (c) Facial expressions

   (d) Tone of voice

10. Which of the following meets the four criteria of a language?

   (a) American Sign Language

   (b) Body language

   (c) Nonverbal expressions

   (d) Gestures

CHAPTER 13

Memory

People are able to respond to environmental events long after the events have passed. This ability is very important for dealing with the world. Psychologists have studied the details of this phenomenon, which we call memory, for many years.

Overview

Memory is one of the most obviously unique features of being human. It is also one of the most remarkable. Years and even decades after an event, people are able to re-create what they have experienced. We can repeat things we have heard. Draw things we have seen. Do again what we have done before. The limits to our re-creations seem to be limited only by our skills, not by the fact that the original events happened long ago. When we sing a song we heard in the past, it may not sound as nice as the original, but that's only because the original was sung by a trained singer with a professional band. Even when we can't re-create, we can identify. When we taste or smell something we have tasted or smelled in the past, we know instantly that we have experienced it before and can say what it is almost as quickly.

> ## Fun Facts: Memory as Show Business
>
> Of all the topics covered in *Psychology Demystified*, memory is the only one that is also a form of entertainment. Folks with extraordinary memory abilities actually perform by demonstrating feats of memory. They recite long lists, answer random questions from telephone books they have seen only once, repeat the names of everyone in the audience, and so on. Unlike so many other psychological phenomena, memory practically leaps out and says, "I'm here! Look at me!"

This ability to bridge gaps in time is not found in natural objects. Plants do not exhibit it. Animals can be shown to exhibit only a very limited form of this ability under very specific circumstances. Some elaborate pieces of technology, such as cameras, audio recorders, and computers bridge time in what seems to be a similar way to humans, but the complexity of these devices just reminds us of how impressive memory is.

It is also important to realize that the capabilities of normal human memory are more impressive than we think. Much of what we think of as remarkable memory abilities are really just people who are putting in the extra effort. A waiter who can remember the orders for a party of ten without writing anything down or making a mistake, or a teacher who can remember each student's name after only one class, is not exhibiting super-human memory abilities. It may be impressive to those of us who can't remember where we left our car keys, but both tasks are well within normal human memory capabilities. We might not put a lot of effort into remembering, but there is a lot of information that we can remember when we do put in the effort or use simple tricks and techniques when memorizing.

Learning and memory are very closely related. In fact, some research can be thought of as either being about learning or about memory, depending on your point of view. We can think of learning as the process of acquiring new information or skills and memory as the retention of what we have learned and its retrieval for future use. Learning and memory work together. You cannot learn if you are unable to remember, and unless you learn you have nothing to remember.

HISTORY

The study of memory is about as old as scientific psychology. Experiments in memorizing lists of words in the late nineteenth century demonstrated some of the first replicable psychological phenomena known. Memory experiments were one of the very few psychological studies that survived the behaviorist revolution almost intact. Behaviorists, always cautious about behavior found only in humans, tended to limit their explorations of memory in ways that kept it separate from the more complex aspects of language. For example, they replaced word lists with lists of nonsense syllables.

One of the pre-eminent features of the cognitivist revolution was a real expansion of the investigation of memory: new experimental techniques; experiments using complex materials, such as sentences and even stories; and an enormous expansion in theorizing.

In short, almost alone in psychology, the study of memory has steadily grown in the number of methods used, phenomena investigated, effects discovered, and theories explored.

CURRENT APPROACHES

Contemporary memory researchers use the information-processing model of human memory derived from the cognitivist revolution. This model argues that memory is analogous to the memory system of a computer. Information enters the system, is coded and processed, stored, and later retrieved.

The computational approach differs from pre-behavioristic views in that the notions of the various information processing capabilities, while corresponding to older notions of storage and retrieval, draw from specific techniques known to work in computers to examine *how* memory works. It differs from behaviorist approaches, which relied on the notion of associations, in suggesting that memory is an active, rather than a passive process.

Memory Systems

The contemporary, computational approach distinguishes several types of memory: *sensory memory, short-term memory (STM),* and *long-term memory (LTM).* Information from the world enters our senses and is briefly retained in sensory memory. If that information is worth attending to, we will code and process it through short-term memory. Over time, with sufficient short-term processing, it may be stored in long-term memory. In that case, it is available days, months, and even years later.

SENSORY MEMORY

Sensory memory is the momentary lingering of sensory information after a stimulus has been removed. A stimulus from the outside world triggers a sensation through one of our five senses. This information is stored in sensory memory for a brief moment before being either further processed or lost forever. For example,

Key Point: Differences in Types of Memory

The differences between sensory, short-term, and long-term memory were discovered experimentally by cognitivists at the dawn of their revolution. Although these separate processes have long been central features in cognitive theory, there is growing direct empirical evidence of their separate existence and different characteristics in brain studies.

if you touch the palm of your hand with the point of your pen, you continue to feel the sensation for a fraction of a second after the pen is withdrawn. If the memory is so fleeting, why is sensory memory necessary? Our sensory memory enables us to start the process of identifying stimuli in order to give it meaning. This marks the transition between sensation and perception.

It is also important to note that we have a separate sensory memory for each of our five sensory systems. We are capable of processing information from our visual system, tactile system, auditory system, and so on simultaneously. Thus, while each individual sensory memory may last only a fraction of a second, we can still store a great deal of information. This does not mean that each piece of information in sensory memory is worth transferring to short-term memory. If we tried to attend to each bit of information that bombarded our senses, we would be overwhelmed to the point of not being able to function. Some kind of selectivity is necessary to weed out the irrelevant sensory information. We will discuss this selectivity later in this chapter. (See the section on "Attention.")

SHORT-TERM MEMORY

Once we have decided that a sensory memory is relevant, that information is processed in short-term memory. Presumably, our short-term memory contains the contents of our conscious awareness, or what we are thinking about at any particular time. The capacity of STM is relatively small, consisting of five to nine pieces of information, called *chunks*.

Both the size and the meaningfulness of a chunk can vary. We can think of STM as having, on average, seven slots in which to put information. You can put a single digit in each slot and process an entire phone number. You could also put an entire phone number in each slot and process seven phone numbers. Thus, the size of the chunks of information varies, while overall capacity does not vary. With further processing, you might replace the telephone numbers with the seven names of the people reached at those numbers in order to be able to differentiate between the business and personal numbers. More meaningful stimuli can be classified in more ways.

Retention of information in STM is also short, lasting approximately 20–30 seconds. If the information is not repeated or processed into long-term memory, the information is lost.

LONG-TERM MEMORY

Long-term memory is really what makes learning and intelligence possible. Because of LTM, our experiences are not lost the moment we stop thinking about them (the moment they leave STM). As a repository of all of our knowledge, LTM contains an extraordinary amount of information, most of which we are not aware of until we have a reason to retrieve it. Unlike STM, both the capacity and the duration of LTM are large. We can retrieve an apparently unlimited amount of information, and many memories can last a lifetime. LTM also enables higher mental functions in another way. It is the availability of these enormous amounts of information for retrieval and association with what we are thinking about that we use to establish meaning.

Psychologists have found it useful to categorize LTM into different subsystems, each representing a specific type of information. The following list describes these memory dichotomies:

- **Implicit vs. explicit memory**   Memory that is expressed in behavior but not consciously brought to mind vs. memory that is deliberately brought to mind.

- **Semantic vs. episodic memory**   Memory of word meaning, general facts, and concepts vs. memory of events that occurred.

- **Declarative vs. procedural memory**   Memory of facts or events vs. memory of how to do things.

Reading Rules

Unlike the distinctions between sensory, short-term, and long-term memory, the evidence that these dichotomies actually involve different brain systems is not so clear. To a great degree, these differences represent differences in experimental procedures and theoretical assumptions. For example, how would you distinguish an implicit, procedural memory from the notion of learning how to do something? Ultimately, combined brain and behavioral research may be the key to determining if there are really separate brain systems or just people thinking in different ways.

Memory Processes

In the previous section, we discussed the three types of memory systems. Here, we will discuss the process of moving a piece of information from one memory system to the next. The process begins by *attending* to a piece of information stored in sensory memory. Next, we *encode* the information into a meaningful form in short-term memory before storing the information in long-term memory. Finally, we *retrieve* the information from long-term memory for use. Each process will be discussed as it applies to either short-term or long-term memory or both.

ATTENTION

As stated previously, not all of the information that enters sensory memory is worth placing in short-term memory. Some kind of selectivity is needed to weed out the irrelevant information. This *selective attention* is the process of restricting focus. For example, imagine that you are at a party talking to a friend. In the background are sounds of other people talking, clinking glasses, music, and other party noises. However, you are still able to follow your friend's conversation. You do this by screening out some of the information entering a particular sensory channel and focusing only on a part of it.

Both the background noise and the conversation are registered by the auditory system and briefly available in auditory sensory memory. You screen out the background party noises and selectively attend to the conversation. You do not, however, completely block out the background noise. These peripheral sounds are given an elementary form of attention. The "cocktail party" phenomenon demonstrates this point. Even when you are in the middle of a loud party, as just described, you can still hear someone call your name from across the room. Thus, you are at least peripherally attending to the background noise for important information.

ENCODING

Encoding is the process of transforming a piece of information into a form that can be stored in memory. *Encoding techniques* are used to transfer information from sensory memory into short-term memory. *Rehearsal strategies* are used to transfer information from short-term memory into long-term memory.

Short-term Memory

Short-term memory encoding is the act of converting sensory stimuli into a form that can be stored temporarily in short-term memory. Existing knowledge or information

is often used to analyze or manipulate the new information. The types of encoding techniques differ on whether or not they are automatic and how deeply the information is processed.

Encoding information from sensory memory into short-term memory can be either effortful or automatic. With some information, we deliberately attempt to transform the information so it is available in short-term memory. For example, you've just met an attractive potential dating partner at a party. You want to get his phone number before he leaves, but no one has a pen. If your solution is to sing the number to yourself using a popular tune, then you've engaged in an effortful encoding strategy. However, some types of information seem to be automatically encoded in memory. We automatically encode information about our location in time or space and how often we experience a specific stimulus. For example, ask yourself what time of day it was the last time you went to a movie, or how many movies you've seen in the last six months. You can probably answer both questions fairly accurately, even though you did not try to remember that information as it happened.

An even more fascinating phenomenon is that we are able to *learn* to automatically encode certain types of information. For example, children learning to read have to work hard to encode written words into short-term memory. Their reading pace is quite slow. They sound out the words and keep them in short-term memory until they can be linked with the word's meaning. Once they've sounded out all of the words, they often have to re-read the entire sentence in order to understand its full meaning. With years of reading practice, this process is not only effortless, it is unavoidable. Try to *not* read a word written in your native language.

Encoding strategies can also differ in how deeply the incoming information is processed. *Structural encoding* is a relatively shallow processing that emphasizes the physical structure of the stimuli. With written words, you would process how they are printed or how many letters are in each word. *Phonemic encoding* is a relatively deeper processing that emphasizes what the word sounds like when spoken. This involves naming or saying the word or recognizing other words that rhyme with it. Finally, *semantic encoding* is the deepest level of processing. It emphasizes the meaning of the word and involves associating objects or actions that the word represents. In general, the deeper something is processed, the more likely the information will be available in short-term memory.

Long-term Memory

Long-term memory encoding involves transferring information from short-term memory into long-term memory. The way information enters long-term memory is not completely understood. However, we do know that the process depends partly on the amount of time we rehearse the information in short-term memory. The longer we rehearse something, the more likely it is to be stored in long-term memory.

Maintenance rehearsal is simply repeating something to ourselves without giving it much additional thought. Repeating the phone number 555-1492 to yourself, either silently or out loud, is a form of maintenance rehearsal. In this case, you are simply re-exposing yourself to the information until you can write it down or store it in long-term memory. If neither happens, the information is lost until you re-expose yourself to it again. Information that is encoded using maintenance rehearsal seldom becomes a part of long-term memory.

Elaborative rehearsal involves taking a new piece of information and mentally doing something to it. This something, or elaboration, can be forming a mental image of the information, applying the information to a problem, or relating the information to something you already know. To continue the phone number example, is there anything familiar about the phone number 555-1492? Can you relate it to something you already know? If you recognized 1492 as the year Columbus first arrived in America, then you have tied part of the new information to something you already knew. Information processed via this type of rehearsal has a much greater likelihood of becoming a part of long-term memory.

RETRIEVAL

Information retrieval is the process of making a piece of information available for use. Short-term memory retrieval involves keeping the information active in memory. Long-term memory retrieval involves accessing the information through recall or recognition and bringing it into short-term memory.

Short-term Memory

As noted previously, the key to encoding information from short-term memory into long-term memory is rehearsing it in short-term memory. The information must be kept active in short-term memory in order to be transferred to long-term memory. We've addressed the types of rehearsal. Here, we'll address the factors that affect the duration of information in short-term memory.

Information only stays in short-term memory for an average of 30 seconds without any rehearsal. However, several factors can either increase or decrease this duration. Recall the example of the phone number, 555-1492, of the potential dating partner you just met. As you are walking across the room to find a pen to write it down, a friend stops you to ask a question. You instantly forget the phone number because your friend interfered with the information active in memory. Thus, *interference* decreases the duration of information in short-term memory.

Two factors can inhibit the effect of interference. First, if the information in short-term memory is linked to information in long-term memory, the information in short-term memory can survive the interference. If you linked the 1492 of the phone number to the year Columbus first arrived in the Americas, then you will be able to retrieve the information from long-term memory and reactivate the phone number (although you may forget the 555 prefix). Second, the motivation to remember the information can also decrease the effect of interference. Increased motivation increases the likelihood that the information in short-term memory is resistant to the effect of interference. Thus, if the potential dating partner is extremely attractive, you will be more likely to keep the phone number active in short-term memory, regardless of your friend's interference.

Long-term Memory

Retrieval of information from long-term memory involves the processes of *recognition* and *recall*. In common speech, we often use these terms interchangeably. In psychology, they represent two very distinct processes. Recognition involves deciding whether or not you have ever encountered a particular stimulus. It is essentially little more than a matching technique. When we are asked to recognize something, we consider its features and decide whether they match those of a stimulus that is already stored in memory. In doing so, we tend to evaluate not the object as a whole, but its various parts. If all of the parts match, the object is quickly recognized. If only some of the parts match, we are left with a vague sense of familiarity.

Recall involves retrieving specific pieces of information, usually guided by retrieval cues. It involves more mental operations than recognition. When we try to recall something, we must first search through long-term memory to find the appropriate information. Then we must determine whether the information we come up with matches the correct response. If we think it does, we give the answer. If we think it doesn't, we search again. The key to finding the appropriate information is to use the correct retrieval cues. These are the cues that we use to help us recall information, specifically during the search phase.

The "tip of the tongue" phenomenon shows the importance of retrieval cues. This phenomenon occurs when the information we are seeking seems almost in reach, but we can't quite get it. This phenomenon is due to the incorrect use of retrieval cues. We are using the wrong, but close to correct, cues to recall the information. For example, you are taking an exam and are asked about a particular definition, but you can't quite come up with the correct answer. You recall the page in the text that contained the definition and that there was a graph on the opposite page, but you can't quite recall the definition. This partial information failed to serve as good retrieval cues for the definition.

One way to avoid this problem is to employ encoding specificity. Your memory for a piece of information will be more reliable if you use the same retrieval cues

Variant View: Schemas, Scripts, and Frames

Traditionally, memory has been studied as we are discussing it here. The issues are about individual pieces of information getting into memory and getting out again. The context in which things are remembered and recalled (or recognized) are studied in terms of their effects on improving or degrading memory. However, over the last 30 years, a different way of looking at context has emerged. Researchers in *schemas, scripts*, and *frames* study how context is reflected in the structures where things are stored in long-term memory.

Outside of school, we rarely try to memorize lists of facts. We remember funny stories, the names of our family and friends, where we keep the frying pan, and so on. There is very good evidence that when we are exposed to facts in context, they are stored in structures that reflect that context. We appear to store stories in standard templates with different slots for the things found in stories: heroes, villains, relationships, fights, rescues, and so on. The structures for stories, called scripts, reflect the plot. (Scripts are not just used for fictional stories. They help us understand and remember news stories and descriptions of events in our lives.)

We also have structures for common places and events. We know that homes have various sorts of rooms: living rooms, kitchens, bathrooms, bedrooms, and so on. We also know that different kinds of furniture are stored in different types of rooms. We know about other sorts of content as well. When we are trying to find the frying pan, we look in the kitchen first. The kitchen is the usual place to find a frying pan. It serves as the *default* location. The structures that identify what goes where are called frames.

Tips on Terms: Flashbulb Memories

Emotion can provide additional retrieval cues for encoding and retrieving long-term memory. Extreme emotions can cause what psychologists call *flashbulb memories*. These are unusually vivid and accurate memories of a dramatic event. It is as though your brain took a photograph of the event and stored a perfect record. You remember exactly where you were, what you were doing, what you were wearing, and so on, when you learned of the event. You can recall this information with perfect clarity even decades later. Some common flashbulb memories in the United States are the assassination of President Kennedy in 1963, the explosion of the space shuttle Challenger in 1986, the fall of the Berlin Wall in 1989, and the destruction of the World Trade Center in 2001.

when you try to recall the information as you used when you stored the memory. Thus, in trying to recall a definition, you will probably have an easier time if you use the same example used in class than you would using a new example. However, this does not mean that you will understand the definition any better.

Memory Failures

Not everything we remember actually happened, and we do not remember everything that happened. Psychologists often learn about normal memory processes by investigating how memory fails. This section describes various types of memory errors.

FORGETTING

If our memory abilities are so vast, why do we forget things that we want to remember? There is an ongoing debate in psychology concerning whether or not information in long-term memory is actually lost or is just not being retrieved correctly. Various theories of forgetting reflect this debate.

The theory of *ineffective encoding* attempts to explain forgetting during the encoding process. According to these theorists, much of what we think of as forgetting is really just a case of failing to encode the information properly. This "pseudo-forgetting" is caused by either the failure to encode the information at all or by using an ineffective encoding strategy. The *decay* theory considers forgetting during the storage process. These theorists argue that forgetting occurs because the physiological memory traces fade over time. Researchers have been able to demonstrate decay in sensory memory and short-term memory, but not yet in long-term memory. Finally, *interference theory* considers forgetting during the retrieval process. These theorists argue that we forget information because of competition from other material. As the amount or similarity of the competing information increases, so does the forgetting.

Retroactive interference occurs when new information impairs the retention of older material. If you study anthropology on Monday for an exam on Friday, and then you study psychology on Tuesday, the psychology information can interfere with the anthropology information during the exam. *Proactive* interference occurs when old information impairs the retention of new material. If you study psychology on Monday and then study anthropology for the Friday exam on Tuesday, the psychology information interferes with your ability to retrieve the anthropology information during the exam.

REALITY MONITORING

We have memories of events that happened to us in the external world, as well as memories of our internal imaginings. Reality monitoring is the process we use to distinguish memories of real events versus imagined events. We are not always accurate. We can often confuse an imagined event as a real one or confuse a real event for an imagined one. For example, a writer imagined a scene in which his main character and his wife were arguing about the wife breaking the TV remote control. A few weeks later, the author's wife asked him where the remote control was, and he replied, "You broke it, remember?" The author makes a reality monitoring confusion by confusing his internal imaginings with events that occurred in the external world.

AMNESIA

Amnesia is the loss of memories over a specific time span, typically resulting from brain damage caused by accident, stroke, or disease. The incident that caused the amnesia is the focal point, and memories are lost for a time span either before or after the incident. *Retrograde* amnesia is the disruption of memories before the incident. A patient will not remember what happened for a time period before, and possibly during, the incident. *Anterograde* amnesia is the disruption of memories after the incident. In this case, the patient can remember what happened before the incident, but will not be able to create new memories for a period of time after the incident.

REPRESSED MEMORIES

The existence of repressed memories is one of the most controversial topics in memory research. Repressed memories are memories of events that have been retrieved for the first time long after the event occurred. They are suddenly recollected and are often traumatic in nature. Most reported repressed memories involve very specific childhood traumas, such as sexual abuse. It is extremely important to determine the truth of such repressed memories because the consequences for everyone involved are so important. However, because the events usually happened years or decades before the memory surfaced, they are extremely difficult to investigate.

There are two major types of research that bear on the question of repressed memories. The first has to do with constructed memories. Research has shown that people remember the gist or meaning of things, rather than recording them literally. When we hear a story that doesn't make sense, we fill in the blanks to make it make sense. A sensible, meaningful story is one we can remember. In real life, people

often experience events very differently than they hear stories about events. In real life, information is missing, we find out things in the wrong order, we only see things from one perspective, and so on.

In the process of making sense of events, we construct a meaningful story and that story is what we remember later on. The story always contains more than the bare facts. Constructed memories are a very ordinary process. But what happens under extraordinary circumstances? What happens when the original events are very traumatic, or we have a critical need to recall parts of the story we never encoded in the first place?

Young children understand less than adults and therefore remember less. As an adult, that same person may have important reasons to remember what happened. Under these circumstances, an adult may rely on construction to a much greater degree than usual. An entire story, complete with details of a traumatic incident that never happened, may be constructed from a few fleeting memories of harmless events.

The second line of research has to do with the practices of the therapists who assist people in recalling traumatic memories. Some of these techniques actually suggest ways to fill in the blank spots in our memories. If the therapist suspects a case of child abuse, she may, even without knowing it, lead the patient to construct a memory of child abuse. Clinically, a very large proportion of the people who report the recovery of repressed memories of traumatic childhood events have recovered those memories for the first time in therapy. While this does not mean the memories are false, it does mean that the therapist may have influenced what was recovered.

The Physiology of Memory

With new technologies, such as brain scanning, our understanding of how the brain stores memories is expanding rapidly, but it is still very limited. Prior to the advent of the new technology, the main way researchers found out what part of the brain did what was to selectively destroy parts of the brains of experimental animals or to study humans who had localized brain damage due to illness or injury.

These early studies identified a part of the brain called the *hippocampus* as being critical for the formation of long-term memories. A famous patient, known as HM, whose hippocampus was removed for medical reasons in the early 1950s is now an old man who wakes up every morning thinking he is seventeen and that Eisenhower is still President. He has no problems with his short-term memory.

The mechanism for storing memories is still unknown. However, a process called *long-term potentiation (LTP)*, which is also implicated in classical conditioning, seems to be involved. LTP occurs when two connected nerve cells are both active at the same time. Repeated activation on both sides of the connection causes the

connection to improve. These changes in connection strength are believed to underlie the storage of long-term memories.

Improving Memory

Even though normal human memory abilities are impressive, there is always room for improvement. Most memory enhancement strategies are based on encoding the new information with useful retrieval cues. At retrieval, we use those cues to recall the desired information. There are too many strategies to cover here, but we will discuss the overarching themes of how they work.

Mnemonic devices are simple techniques that can greatly improve anyone's powers of recall. Many of these internal devices involve clever ways of organizing material when it is stored in long-term memory. The *method of loci* is one such technique. This method involves associating a list of items to be remembered with a series of locations (or loci) that are already firmly fixed in memory. If you wanted to remember your grocery list, you'd visualize each item on the list in a location on a path you know very well, such as your route from home to work.

Visual imagery is a technique that involves relating verbal information to some kind of visual image. The waiter discussed at the beginning of the chapter might use visual imagery to remember the *diners'* orders. The woman who ordered chicken may have a bird-like nose, and the man who ordered shrimp may be short. The association between the food order and a characteristic of the diner may help the waiter remember not only what is ordered, but who ordered what.

A third technique is based on the internal environment of the learner. *State-dependent* memory involves using internal cues, such as mood or health, to recall information. For example, if you study for an exam while you are depressed, excited, or suffering a cold, you are more likely to recall the information if you are in the same state as when you learned the information. Therefore, if you are studying while you have the flu, then you may be better off if you take the test while you have the flu (although your classmates or professor might *not* appreciate your being there). On the other hand, you can also study for the exam when you feel better and take the test when you are healthy.

A fourth technique is based on the external environment of the learner. *Context-dependent* memory involves using external cues, such as pictures on the wall or your location in the room, that aid in the retrieval of information. The best place to study for an exam is in the exact place where you will take the exam. Conversely, the best place to take an exam is in the place where you studied the information. The point is that there are cues in your surroundings that can help you remember information. If you move, then the cues are slightly different or not present at all.

The overarching theme to each of the enhancement techniques discussed so far is to relate the new information to something you already know. Adding meaningfulness to memory increases the likelihood of later recall because the old information acts as a powerful retrieval cue for the new information. One way to attach meaningfulness to memories is to process the new information in relation to a *schema*, an organized cluster of knowledge about a particular object.

A schema is helpful for memory because it provides organization and is usually related to other information stored in long-term memory, which aids in retrieval. For example, if you're studying the terms and concepts of Pavlovian conditioning, it helps to process the information within the schema of Pavlov's experiment. When you are asked, perhaps on an exam, to identify the conditioned stimulus in Pavlov's experiment on conditioned salivation, you can search your Pavlov schema to answer the question.

Summary

In an odd sense, the study of memory is too easy. This is not to say that many sophisticated experiments and complex theories have not been developed. In fact, it is the ready availability of so many memorial phenomena that has allowed psychologists to make such enormous advances in studying memory. It is very easy to find all sorts of situations where people demonstrate the effects of past experience. It is extraordinarily difficult to imagine how these effects might be caused by something *other* than a brain mechanism that works like a mark on a cave wall, a filing cabinet, or a computer disk. These sorts of metaphors are central to almost every theoretical explanation for almost all of the empirical evidence of memory. But, the empirical evidence only shows that people are affected by their past experience in these particular ways. There is very little evidence that the brain is organized like a cave wall, a filing cabinet, or a computer disk. Psychologists who study memory will have to wait for brain science to verify their theories.

Quiz

1. Which model argues that memory is analogous to the memory system of a computer?

    (a) Behaviorists

    (b) Information processing

(c) Freudian

(d) All of the above

2. Which of the following is not considered to be one of the types of memory stores?

   (a) Sensory memory

   (b) Short-term memory

   (c) Flashbulb memory

   (d) Long-term memory

3. _____ occurs when new information impairs the retention of older material.

   (a) Implicit interference

   (b) Explicit interference

   (c) Proactive interference

   (d) Retroactive interference

4. Which type of memory contains the contents of our conscious awareness, or what we are thinking about at any particular time?

   (a) Sensory memory

   (b) Short-term memory

   (c) Long-term memory

   (d) Episodic memory

5. Which type of memory can be considered the repository of all of our knowledge?

   (a) Sensory memory

   (b) Short-term memory

   (c) Long-term memory

   (d) Episodic memory

6. _____ memory is expressed in behavior but not consciously brought to mind?

   (a) Implicit

   (b) Explicit

   (c) Semantic

   (d) Episodic

7. The first step of committing something to memory is to _____ the information.

    (a) Encode

    (b) Retrieve

    (c) Memorize

    (d) Attend

8. Interference _____ the duration of information in short-term memory.

    (a) Decreases

    (b) Increases

    (c) Maintains

    (d) Does not effect

9. Which of the following is not a theory of forgetting?

    (a) Decay

    (b) Ineffective encoding

    (c) Interference

    (d) None of the above

10. Which area of the brain is believed to be responsible for forming long-term memories?

    (a) Hippocampus

    (b) Limbic system

    (c) Pons

    (d) Corpus callosum

CHAPTER 14

Cognitive Psychology: The Study of Thinking

The word *psychology* literally means the study of the mind. The mind is presumably what people have that allows us to do what animals, vegetables, and minerals can't do. And what we can do that those other things can't do is think. But what is thinking exactly? Cognitive psychology is the branch of psychology devoted to answering this question.

Overview

Let us try to characterize thinking in the broadest terms: A person is confronted with a difficult situation. He doesn't know what to do. He thinks for a while, and then he does know what to do. Thinking is a typically invisible, undetectable process that enables people to respond effectively to difficult and complicated situations where simple reactions generally will not suffice.

Doing science about something we cannot see is hard. An additional problem is that, although we can't detect or observe *other* people thinking, to some extent we can observe ourselves thinking. We might try to study thinking by just asking our subjects

how they think about something. The problem here is that science requires that we be able to *verify* our observations. If we ask people to tell us what and how they are thinking, how do we check to see if they are right? So far, psychology has found no answer to this question. As a result, cognitive psychologists study thinking indirectly.

HISTORY

What people are thinking about is a question for philosophers, sociologists, social and developmental and clinical psychologists, political scientists, novelists and playwrights, and a host of other people. Cognitive psychologists are more concerned with *how* people think about what it is they think about than with the details of what people think about. This is a concern they share with logicians, who have been dealing with these problems for a lot longer than psychologists have even been around.

Aristotle (384–322 B.C.) was the first Westerner to worry about how people think. He developed two important distinctions that are still in use today. The first is the distinction between form and content. Aristotle noticed that there are some very common words, such as *and, or, if, then, all, some, none*, and a few others, that tell us about the relation between ideas. In the right context, these special words, called *logical connectives*, can tell us the right answer even if we don't know the details of what is being talked about. For example, suppose you had a bad telephone connection and heard someone say, "It was either P or Q, and it wasn't P!" where P and Q were so garbled that you couldn't tell what was being said. You would still know that, whatever Q is, that is what happened.

The *content* of our speech and our thinking is just a name for whatever it is we are talking or thinking about. In this example, we don't know what the speaker is talking about, so we just wrote down P and Q to stand in for it. We don't know the content. All we know is the *form*. The form is the structure of the speech or thinking, usually using those special logical connectives, but leaving out the content.

The importance of the distinction between form and content is that it means that, sometimes, we can know we are correct or incorrect without knowing what we are talking about. Much of *how* we think is just about form, not content. Logic is the study of how to distinguish between correct and incorrect thinking on the basis of form alone.

Key Point: A Benefit of Logic

An important benefit to being able to distinguish form from content is that most areas of disagreement come about only when content matters. If the correct answer can be calculated without knowing the content, then the argument can be settled by relying on logic alone. In modern applications, however, where complex mathematics has been added to the logic, people can still disagree as to whether the right sort of logic is being used.

Aristotle also distinguished between two types of thinking: practical reasoning, where we worry about when to get up and what to buy at the grocery store and what road to take to work, etc., and pure reasoning, where we think abstractly about truth and beauty and politics and literature and art. (Aristotle was probably a bit of a snob.)

The distinction between practical and pure reason is still important, although we think about it differently these days. For example, the twentieth century philosopher, Gilbert Ryle, pointed out that there is a difference between *knowing how* and *knowing that*. We could read a book on bicycles and learn all about the mechanics and the engineering and even the principles of how to ride and still not know how to ride a bike. We would fall off (at least the first time). On the other hand, some people who can't read can ride bikes. Practical know-how and pure, intellectual knowledge do seem to be separate. Modern brain studies back this up.

Ryle also pointed out that his distinction also applied in the abstract world of pure philosophical thought. Literary critics who understand deeply and profoundly what makes for good writing and even write books explaining how to write well often write terribly, even while they are explaining how to write well. And there are novelists who have no idea about literary theory and couldn't tell the difference between great writing and horrible writing but, nonetheless, are great writers themselves. In other words, the skills for doing things correctly are separate from the skills for understanding how to do things correctly.

For several hundred years after Aristotle, logic made great progress. After that, things slowed down for a long while. The problem turned out to be that logic can get very complicated very quickly. Without good tools, it is hard to keep everything straight. In the middle 1800s, a logician named Boole figured out a way to use symbols (like the Ps and Qs discussed previously) to work with the forms of sentences. Following the example of algebra, he developed rules for combining various logical operations so that, even with very complicated arguments, the correct answer could be calculated. This system was the first logical calculus, called the *propositional calculus*. Computers work using a system based on Boole's propositional calculus.

Logic began to make tremendous progress again, a progress that continues to this day. About 30 years after the development of the propositional calculus, other, more powerful systems began to be developed. Problems could be worked out on paper that could not be solved just by thinking. It became clear that there was another important distinction to be made in the study of thinking. Logic could tell us how to think correctly, but it couldn't always explain how we actually think. People had been thinking for tens of thousands of years without symbolic logic. They must have been making mistakes, but they still got by okay. The new symbolic logic could detect our mistakes once we knew exactly what people were thinking, but it couldn't explain how we were getting by with less than perfect logic.

Psychologists distinguish between *descriptive* and *prescriptive* accounts of thinking. Logicians study what sort of thinking is correct, the sort of thinking people should do, the sort of thinking we might prescribe. Psychologists take their job to be the study of what sorts of thinking people actually do. Cognitive psychologists describe thinking.

The first psychologists, called *structuralists,* relied upon the technique we mentioned at the beginning of the chapter. They asked subjects what they were thinking. This technique was called the method of *introspection.* The subject "looked inside" and reported what he saw. Because of the unreliability of this method, introspection is rarely used today and only for specific purposes.

CURRENT APPROACHES

We cannot rely on people to report on their own thinking, not because we don't trust them, but because we have no way of checking. Even if we are perfectly honest and make no mistakes, there are very likely things about our own thinking that we can't detect. For big complicated problems with many steps, you can probably tell someone else what steps you took in what order. But what if I ask you a simple question, like, "What is two plus two?" or "What was the color of the house you grew up in?" How much could you tell me about what you actually did to come up with the answer?

Cognitive psychologists have developed a number of methods for working around this problem. The first method relies on logic and related systems. Cognitive psychologists give tasks to their subjects where the correct answer is already known. They measure the subjects' responses by comparing them to the right answers. The pattern of wrong answers can often tell us a lot about how people are thinking about things. For example, one theory might predict that certain problems are harder and another theory might say that other problems are harder. Depending on which types of problems people get wrong more often, we might be able to distinguish between the two theories. (We will see examples of this sort of research later on in the chapter.)

Another common way of finding out how people are thinking is to measure the time they take to perform a task. These measures are called *reaction times.* For example, complex cognitive tasks take a lot of steps to work out and that takes time. Logic may show us that there are shortcuts available that still give the same answer. By measuring the time that different people take to perform different tasks, cognitive psychologists can theorize about what methods people are using to think about those tasks.

Psychologists actually began to study thinking using these methods over a hundred years ago. However, for the first half of the twentieth century, little research was done because the behaviorists, who had abolished introspective research, were suspicious of studying mysterious, internal behavior like thinking.

Tips on Terms: Heuristics

Oddly enough, logic can show us that there are certain methods for solving various types of problems that don't work all the time. Some of these methods, called *heuristics,* are quick shortcuts that work enough of the time that they may be very useful in day-to-day thinking. The problems people encounter from the occasional wrong answer are offset by the time and energy they save by using the heuristic to get a quick answer.

Using methods such as logic and reaction times, cognitive psychologists can determine whether people are using heuristics and which heuristics they are using.

Then, in the 1950s, came the *cognitive revolution.* As you might imagine, the cognitive revolution had something to do with cognitive psychology. In psychology, the cognitive revolution came about when some very clever psychologists began to put together combinations of methods, such as logic and reaction times, to figure out indirectly how people think. These psychologists saw the cognitive revolution as freeing them up from the unnecessarily restrictive rules enforced by the behaviorists.

But the cognitive revolution had another side as well, a theoretical side. Not only psychologists, but also philosophers and linguists and computer scientists and others began to think about thinking in terms of some of the new advances in logic and related mathematical theory that had come along in recent years. The most important new theory was about the concept of *information.* Information theory showed how the logical structure of a message can differ between the sender and the recipient. But the key to this new theory was a new way of measuring the amount of information at one place or the other. The more information we have about something, the more possibilities are eliminated. For example, if someone tells you there is a pair of shoes in the other room, you still don't know the color, or whether the shoes are high heels or sneakers, or whether they're new or old, etc. If, however, someone tells you his old green sneakers are in the other room, he's given you more information and you can eliminate more possibilities.

This notion of measuring the amount of information gives cognitive psychologists a third way of measuring indirectly how people think. Cognitive psychologists are able to set up experiments where the likelihoods of the various possibilities are known, at least approximately. Depending on how much of the available information the subjects are using, they will eliminate more or fewer possibilities as they think about the task. Because the psychologists know the exact likelihoods, they are able to translate the possibilities into probabilities or odds. Depending on the proportions of answers given to different questions or on the subjects' own estimates of the likelihoods of various events, the psychologist can figure out directly how much information the subjects are using. If one theory says that people use more

Big Background: Cognitivism

There is another, more philosophical side to the cognitive revolution, called *cognitivism*. Cognitivism greatly influences how cognitive psychologists (and other psychologists, including some developmental, social, and even clinical psychologists) do their research. The idea behind cognitivism is that the human mind works like a computer in at least two important respects. First, the processes of sensation and perception result in representations (discussed in Chapter 12, "Psycholinguistics: The Psychology of Language"). In other words, brain states act like linguistic symbols. Second, the processes, including thinking, that result in behavior operate on these representational symbols according to the same sorts of rules that computers use for manipulating symbols.

As we will see, the theories of various sorts of thinking put together by cognitive psychologists are deeply influenced by this computer metaphor. Cognitive psychologists see the process of thinking as literally being a "process," with steps and stages and calculations, very much like the way a computer works.

---

information than another theory does, then these measurements can help tell us how people are thinking, similarly to how logic and reaction times do.

In short, cognitive psychologists have developed elaborate means for testing how people think by setting up various tasks and seeing whether subjects get the right or the wrong answer, how long they take to get an answer, and how likely they think various events are to occur. In the remainder of this chapter, we will see how these methods are used to unravel the mysteries of human thinking.

Organizing Information

Before we look at how we combine information to solve problems, we have to understand how we organize that information. If you were asked what you see in front of you at this moment, you would probably answer with a series of one- or two-word labels (book, coffee mug, television, and so on). This reveals an interesting and important aspect of human cognition. Although the world consists of a multitude of objects and events, each in many ways unique, we tend to simplify and order our surroundings by classifying those things together that are similar in some way. We use *concepts* to categorize things in the world.

We also formulate concepts about how things in the world are related to each other. Consider the two concepts of flame and heat. We have an implicit rule that the former causes the latter. This is a general rule that we can apply to particular situations.

Critical Caution

Cognitive psychologists use the word *information* in two very different ways. There is the formal mathematical definition of information taken from information theory, discussed earlier. Cognitive psychologists also use the term *information* to talk about the meaningful content of the representations used in cognitive processes. It is in this second informal sense that we speak of concept formation and categorization as organizing information for later use.

Thus, when we see a flame, we do not put our hand into it. The ability to form concepts of objects and the relationship between objects allows us to simplify our environment into manageable units.

It is difficult to overstate the importance of concepts to cognitive processing. They are the building blocks of thought. They allow us to impose structure, and therefore predictability, on an environment that would otherwise consist of millions of unrelated stimuli. This is partly done by organizing information into manageable units. With concepts, we take things that are similar to each other, but not exactly the same, and treat them as if they were equivalent. For example, even though two lamps may not be identical, we treat them as members of the same concept and interact with them similarly. We plug them in and turn them on when we need light. It doesn't matter that one is a floor lamp and the other is a desk lamp.

Imagine that humans did not have the ability to form concepts. Every time we encountered a new stimulus, we would have to figure out its purpose, whether or not it was harmful, and how to use it. When faced with a new lamp, we would have to figure it out. Now, consider how many novel stimuli we encounter in an average day. Imagine how much of our time would be spent having to figure out that the computer in the computer lab at school served the same function as our computer at home or that the fork in the restaurant served the same purpose as our fork at home. We would not be able to function, as this process would take up most of our cognitive processing. This is why concepts are important. They allow us to reserve our cognitive processing for other functions by allowing us to treat similar objects as equivalent.

THE STRUCTURE OF CATEGORIZATION

The way we categorize concepts is equivalent to asking the question "how do we think?" The answer to this question is based on a structure or organization that we impose on the world. Researchers have determined that we impose a hierarchical organization on our concepts or categories of objects. By definition, a hierarchy has

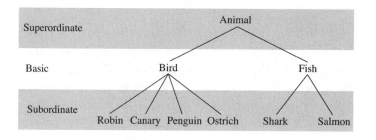

Figure 14-1   Organizational hierarchy

several levels that differ in their level of inclusiveness, or what is considered a member of the category. The level of inclusiveness indicates what objects, or instances of the category, fit in a particular level of the hierarchy.

Figure 14-1 illustrates a simple organizational hierarchy with three levels. The *basic level* is the middle layer in this example. It is the level at which we use and define objects on a daily basis. It is our default level of categorization. This level has moderate inclusiveness, which minimizes the cognitive effort associated with making too many fine discriminations. In our example, "bird" and "fish" are basic level categories. Without the need to convey more or less detailed information, we typically identify animals at this level of inclusiveness.

In order to feed the small animal with feathers and wings in our backyard, we only need to categorize it as a "bird" in order to interact (feed) with it. Categorizing it as an "animal" will not help us to determine whether to put out bird seed or cat food. Categorizing it as a European swallow or an African swallow is more information than we need because both types of birds will eat the same seed. Thus, unless the situation dictates that more or less fine-grained information is necessary, we generally categorize objects at this basic level.

The level below the basic level is the *subordinate level*. It is the least inclusive in that it includes fewer instances within each category. Thus, even though all members of the subordinate level are members of the basic level, not all members of the basic level are included in each subordinate level category. In our example, "shark" and "salmon" are subordinate level categories of "fish." If you are at a restaurant, ordering "fish" will not convey enough information to your waiter about what you want to eat. You could end up with either a nice filet of salmon or a guppy. A more fine-grained level of detail is necessary to achieve your goal. Some occasions dictate that you convey more information than the basic level, and you move to the subordinate level to convey this information.

The level above the basic level is the *superordinate level*. It is the most inclusive in that it includes more instances of the category than the basic level. Thus, while all instances of the basic level are members of the superordinate level, not all instances of the superordinate level are included in each basic level category. In our example, "animal" is the superordinate level. While it is less common for people to need to convey less information than the basic level, such situations do occur. Typically, this occurs when we do not have enough information about the object to categorize it at the basic level. Suppose that you briefly see a small furry object in your backyard. You don't have enough information to determine if it is the neighbor's cat, a skunk, or a squirrel (all basic level categories). To call attention to this object, you can only categorize it as an animal.

THE PROCESS OF CATEGORIZATION

How do we determine whether an object is a member of a category? When we are faced with a novel stimulus, we mentally search for a category in which to place it. This is essentially asking the question, "What is it?" There are two processes by which this occurs. Categorization by *features* occurs when we look at an object and assess what features characterize that object. The features themselves define the category. We compare the new object with our knowledge of the category in memory. If it shares a number of features, we call it a member of the category and treat it as such. If it does not match the category, we continue searching for the appropriate category. You can think of a list of features as a checklist. For example, the three features that define the category "bachelor" are male, adult, and unmarried. You would not categorize a twelve-year-old unmarried male as a bachelor because the age feature does not match.

Another, more typical process is categorization by *prototypes*. A prototype is an example that best represents the category. We compare the new object to a prototype of the category in memory. Thus, this is a matching process. For example, we all have an idea of what a bird looks like. When we are faced with a robin, we can quickly categorize it as a bird because it readily matches our prototype. If we are faced with a penguin, we have a more difficult time categorizing it as a bird because it doesn't match our prototype as well. It takes us more time to properly categorize a penguin than it does to categorize a robin.

The difference between using features and using prototypes is that, with features, there is a strict set of rules for a match. Every rule must be satisfied. If the answer to one question on our checklist is wrong (like the age of the twelve-year-old unmarried male), then we do not get a match. With prototypes, the rules are much more flexible. Typically, birds have feathers, fly, build nests, have beaks, and so on. Although the penguin doesn't fly or build nests and it swims underwater, it matches enough items on the checklist (and fails to match enough on checklists for other categories, such as fish) so that we categorize it as a bird. The rules for prototypes are called *soft constraints*.

Problem Solving

Knowledge of concepts or categories, and the relationships among them, make problem solving possible. If you did not understand how the concepts in a problem were related, you would not be able to solve the problem. For example, your goal is to take six matches of the same size and assemble them so that they form four equilateral triangles with every side equal to the length of one match. If you found that the only way to solve this problem was to build a three-dimensional pyramid (really, a tetrahedron), then you had to have knowledge of the concepts of "triangle" and "pyramid" and how they relate to each other.

There are three stages to problem solving. First, we must represent the problem in memory. Second, we must design a solution strategy. Third, we must decide when a solution is satisfactory.

REPRESENTING THE PROBLEM

No single aspect of problem solving has a greater impact on the speed and likelihood of finding a good solution than the way in which a problem is represented in memory. How you think about and interpret the problem has a profound affect on whether or not you successfully solve the problem. For example, what is the length of the hypotenuse h in Figure 14-2?

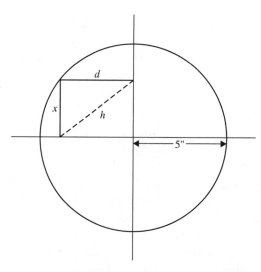

Figure 14-2 Sample problem

The answer is 5 inches. If you tried to solve this problem by trying to remember the formula for calculating the length of a hypotenuse, you are representing the problem in a more difficult way than is necessary. You could represent the problem by drawing an imaginary line from the origin (at the center of the circle) to the intersection of x and d. Your knowledge that the radius from the origin to any point in the circle is constant and that the radius is 5 inches allows you to solve the problem without more complicated calculations. What first appears to be a difficult problem becomes easier when you represent it differently.

There are several types of stumbling blocks we impose on ourselves when solving problems. First, we often try to use all of the available information in a problem, although not all of the information will help us reach a solution. We fall prey to *irrelevant information*. We assume that if the information is included, it must be there for a reason. For example, if 10 percent of the people in your town have unlisted phone numbers, and you randomly select 300 names from the phone book, how many of those will have unlisted phone numbers? If you answered 30, then you focused on the numbers in the problem and included irrelevant information when representing the problem. If you reasoned that none of the 300 numbers would be unlisted, then you ignored the irrelevant information and concluded correctly that no phone numbers in the phone book would be unlisted by definition.

A second problem is that we place *unnecessary constraints* on ourselves when solving a problem. When we interpret a problem quickly, we may be hindering our ability to solve it. Once we have committed to representing a problem in a certain way, we automatically structure the available information accordingly. This reduces our chances of seeing better alternatives. For example, a dog has a 7-foot chain fastened around his neck and his water bowl is 12 feet away. How can the dog reach the bowl? If you are having trouble solving the problem immediately, then you are likely representing the problem with the assumption that the other end of the chain is affixed to some object (a tree, for example). By not making this unnecessary constraint, you could solve this problem easily by positing that the dog just walked over to his water bowl, dragging his chain behind him.

A third problem is that we fall victim to *functional fixedness*. We often perceive objects as having only their typical functions. When an object can be used in a typical manner to solve a problem, we often overlook it. In a classic example, suppose that you are in a dark room and have only a box of matches, tacks, a candle, and a string. You need to arrange these objects to provide you with light, while keeping both of your hands free. In other words, you cannot just hold a lit candle. The solution is to use the match box as a platform for the lit candle. It is secured to the wall by a string tacked to both the box and to the wall. How does functional fixedness hinder a solution? Most people only consider the box of matches to be used to light the candle. Because this is the typical use for matches, most do not consider using the box itself as a platform for the candle.

Finally, we have a tendency to become stuck in a mental set. We use problem-solving strategies that have worked in the past. In most cases, this is fine. However, using the same strategy may blind us to alternative solution strategies that require fewer steps or less cognitive processing. Thus, we may overlook a simpler solution in favor of a more complicated solution that worked in the past. A classic demonstration of the problem of mental set is found in the Luchin's water jar problems. A person is given a series of six problems in which she is given three jars of water (jars A, B and C) of varying volume and told to arrive at a certain volume. The solution to the first five problems is to use the formula $B - A - 2C$. However, the solution to the sixth problem is either $B - A - 2C$ or $A - C$. If people fall pray to mental set, they will overlook the easier solution in favor of the more complicated one.

DESIGNING SOLUTION STRATEGIES

Most problems cannot be solved instantly, even when they are represented in effective ways. We often must go on to manipulate the information in some fashion before the final solution emerges. We use several strategies to manipulate the information contained in a problem.

One method is to apply an *algorithm* to the problem. An algorithm is a precisely stated set of rules for solving problems of a particular kind. The major advantage of this is that it guarantees success if it is applied in the right circumstances and is followed correctly. For example, the formula πr^2 provides an algorithm for finding the radius of a circle, whether that radius is in inches or miles. If the algorithm is applied to calculating the radius of a circle, it will yield a correct solution every time. The disadvantage of using algorithms is that most problems we encounter do not lend themselves to such solutions.

A second method is to design a solution using *trial and error*. This involves trying each possible solution until the correct one is found. This is a good strategy but only when there are a few possible solutions to test. When there are many possible solutions, this method is inefficient because of the time and/or resources it takes.

Finally, we can use *heuristics* to solve problems. Heuristics are general rules of thumb strategies. Unlike algorithms, they do not guarantee a solution, but they frequently pay off with quicker solutions. There are several general types of heuristics that people commonly use. One method is to perform a *subgoal analysis*. This involves breaking down a large unmanageable problem into smaller problems, or subgoals. For example, an average person beginning a chess game could never consider all of the 10^{120} possible play sequences that are theoretically possible. Clearly, chess players have to limit their focus. The strategy is to break down the problem of winning the game into smaller problems, such as protecting the queen or controlling the center. The demands on the player's limited information processing capabilities are substantially reduced. However, that is no guarantee he will spot the best move.

A second heuristic method is to use a *means-end analysis* on the problem. This strategy involves comparing your current position with a desired endpoint and then trying to find a means of closing the gap between the two. This method is slightly biased in that it encourages a focus on reducing the existing distance. If the problem must be solved by first increasing the gap before reducing the gap, this analysis may bias you away from the best solution.

Finally, we can perform a *backward search* strategy. This involves beginning at the endpoint and working your way to the beginning. For example, when trying to solve a maze puzzle, searching backward saves time. There are more lanes that lead away from the starting point than lead away from the endpoint. Thus, searching backward saves both time and effort.

SATISFACTORY SOLUTIONS

How do we decide when a solution is satisfactory? Usually, being satisfied with a solution is easy to determine if it is a question of the solution being correct or incorrect. We are satisfied with correct solutions and unsatisfied with incorrect solutions. However, when determining the best solution for a problem with multiple solutions, this question becomes more difficult. In these cases, we often settle for a less than ideal solution when finding a better solution stresses our cognitive capacity. For example, how do you decide when you have satisfactorily solved the problem of writing a term paper? You may write and rewrite sections of the paper, or put it away while you regain cognitive capacity, or just turn it in because it is due. In each case, we work on the solution until we reach some criterion of satisfaction.

Note that the criteria for a satisfactory solution may resemble the soft constraints of categorization theory. In a sense, we stop the problem-solving task when we categorize the problem as being solved.

Reasoning

Do people think logically? In other words, when people think through problems for which logic says what to do, do they do it that way or not?

Logic does not tell us the correct way to handle every problem. The sorts of problems with which logic can help are word problems where a situation is described and the facts not given in the problem can be figured out without any additional background information. You are probably familiar with these sorts of word problems from the analytical sections of standardized tests. For example, you are told certain facts about a seating arrangement at a holiday dinner and have to figure out where a particular person is seated, even though that isn't specifically mentioned in the problem.

Reasoning is the name we give to the way people use logic to work with word problems, handle political debates, solve mysteries, and so on. So long as both what we know and what we want to know can be expressed logically, the thinking we use can be understood in terms of reasoning and logic can be a help.

Logic isn't only used for these special types of problems. Embedded in ordinary thinking, we often have the information we need, but not stated in a convenient way. Suppose you need to talk to Jane. If we know that all the American History students are in the classroom because there is a test today, and we know that Jane is taking American History this semester, we know that Jane is in that classroom. Logic won't tell you how best to get hold of Jane, or how to tell her what you need to tell her, but it has told you where she is.

LOGICAL ANALYSIS

Logic is designed to tell us how to start with statements about things we know and come up with new statements we didn't know when we started. Logic does this by distinguishing between correct and incorrect *inferences.* An inference is where, starting with two or more facts, called *premises,* a person asserts another fact, called the *conclusion.* When we extract the logical form of an inference, leaving out the content, we are left with a *logical argument.* Logic can tell us if the argument is *valid* or *invalid.* A valid logical argument is where knowing the premises assures us something about the conclusion. Logic gives us a guarantee about the conclusion whenever the form of the inference matches the form of a valid argument. In other words, whenever a person says something new based on things she already knows, we can use logic to see if she is justified in doing so.

The strongest form of inference is the *deductive* inference. When the premises of a deductive argument are true, the conclusion must be true. In other words, if an inference matches the form of a deductively valid argument, knowing that the premises are true justifies believing that the conclusion is true. Take our example from the overview earlier in this chapter. Suppose we know that the winner of the election was either a Republican or a Democrat (because the other candidates didn't have a chance). Then we hear that the Republican didn't win. Without hearing anything else, we can conclude that the Democrat won. The reason we can be confident in our conclusion is because the form of our inference—*(Premise 1) P or Q, (Premise 2) Not P, (Conclusion) Q*—is a deductively valid argument. No matter what P or Q is, we know that if the premises are true, then so is the conclusion.

Variant View: Inductively and Abductively Valid Arguments

Depending on which logician you ask, there are either one or two weaker forms of valid argument as well. Many logicians claim that there are *inductively* valid arguments. In an inductively valid argument, knowing that the premises are true guarantees that the conclusion is probably true. Some logicians claim that there are also *abductively* valid arguments. In an abductively valid argument, knowing that the premises are true guarantees that the conclusion (called a hypothesis) is a possible explanation of the premises. The hypothesis is then considered plausible.

THEORIES OF REASONING

Cognitive psychologists are not only interested in when people make logical errors, they are also interested in what sorts of processes allow us to draw conclusions from premises. How do we make inferences? There are three main theories:

- **Mental logic**   People make inferences by creating logical proofs that mechanically derive conclusions.

- **Mental models**   People make inferences by creating models of the world based on the premises and then check to see if the conclusions are true in all of the models.

- **Schemata**   People make inferences using domain-specific systems to derive conclusions mechanically.

Mental logic is the theory that the cognitive processes work the same way as the logical processes logicians have discovered for deriving logically correct conclusions from premises. (*Derivation* is the name logicians give to the calculation process from premises to conclusions.) The biggest problem with mental logic is that it has trouble explaining how the content of a word problem affects people's reasoning. In experiments, given different word problems about different topics with the exact same logical form, people make different sorts of errors. Because logic deals only with logical form, differences in content shouldn't make a difference.

The mental models approach uses a system derived from logic for testing conclusions. Unlike derivation, this system does not generate conclusions. The mental models system differs from the logical system in that it can take some degree of content into account. One problem with the mental models approach is that it does not explain where subjects get the possible conclusions to test. A second problem is that many logical problems do not immediately lead to a plausible set of models of the world. It is unclear whether mental models are used in the more automatic types of logical thinking.

The schematic approach uses a system closely related to the schemas, scripts, and frames discussed in Chapter 13, "Memory." Different content domains, such as causal relations, rules and their enforcement, or the social exchange of goods and services, each have their own schemata. This provides each domain with its own mini-logic, where derivations can be performed. A problem with this approach is that it is hard to determine how many domains there are and what the rules are for each.

Decision-making

We are faced with many decisions in life. Often, logic alone is not enough to help us make those decisions. Logicians and mathematicians have developed additional tools for decision-making. These tools are particularly useful when we cannot know things for certain, but only to some degree of probability. Inductive, rather than deductive, inferences are needed. Cognitive psychologists study how people actually make decisions to see how their decision-making process differs from what the mathematicians recommend.

DECISION THEORY

Decision theory tries to describe what a "rational" decision-maker would do in any particular choice situation. Rationality is described using a number of assumptions, such as preferring more money to less money, being willing to exchange equal valued objects, and so on. Decision theory calculates mathematically what someone who keeps to these assumptions would do if presented with this or that decision.

Key Point: Math Versus Logic

While the strategy for studying decision-making looks like another case where psychologists look for right and wrong answers to compare to a mathematical theory, the situation is not that simple. Compared to the role of logic in reasoning, decision-making theories require many more assumptions about what is the best way to solve a problem before they can give us an answer. Therefore, the right and wrong answers that the mathematical theory gives are only right or wrong if the assumptions are true. In many of these experiments, the answers that subjects give are not necessarily wrong, just at odds with the theory.

The basic element in decision theory is the alternative. In life, decisions are needed when we face multiple alternatives. Which restaurant to go to for dinner. What movie to see. What elective course to choose. Whether or not to go out on a date. And so on. Decision theory helps us make sensible decisions depending on how much we know about the various alternatives.

Decision theory requires that all alternatives be comparable. These comparisons are made using the possible consequences of choosing each alternative. The consequences are called *outcomes*. Each outcome can be compared to the others because all outcomes are measured in terms of *payoffs*. Payoffs may be measured in money, points, or any other unit that is common to all the alternatives.

Decision theorists distinguish between two types of decisions: *decisions under uncertainty* and *decisions under risk*. In both of these, the alternatives, outcomes, and payoffs must be known in advance. The difference is that, for decisions under risk, the probability of each outcome given the choice of each alternative must also be known.

For example, given that you choose to see the latest Brad Pitt movie (an alternative), there is a certain probability that you will like the movie, that you won't like the movie, that all the tickets will be sold, or that the movie will no longer be showing. If you know the probabilities for each of these four outcomes in advance for each movie alternative, then you can use the theory of decisions under risk. If you don't know the probabilities, you must use one of the theories of decision under uncertainty.

HEURISTICS AND BIASES

Cognitive psychologists usually study decision-making by providing subjects with word problems that describe hypothetical situations calling for a decision. They compare the choices made by the subjects with what various decision theories recommend. Two psychologists, Daniel Kahneman and Amos Tversky, discovered a number of types of problems where subjects consistently behave in ways that decision theory would say are irrational. They argue that people confronted with these decisions use heuristics that work well most of the time in real life, but fail in these carefully constructed examples. These special example problems divide people who use these heuristics from those who make choices using the more cumbersome calculations of decision theory.

Tversky and Kahneman and their colleagues have identified heuristics for decision-making. Two of these are the *availability heuristic* and the *representativeness heuristic*. Like all heuristics, they are quick shortcuts that work well in many settings, but don't work in certain situations.

The availability heuristic is a way of estimating probabilities for events (a necessary step if you want to deal with decisions under risk). The idea is that the more often you have heard about something, the more common it is. Unfortunately, each of our own personal experiences is not necessarily a good statistical sample of the real world. The proportions don't always match up. Worse, in this world of mass media, any bias shown by the media automatically becomes a bias in the experience of hundreds of millions of people.

For example, in America, which event do you think is more common: suicide or murder? Most people say murder because murder gets lots of play in the news and in fictional TV detective shows. Actually, suicide is more common. Intriguingly, our individual experience can be biased while our collective experience can still be reliable. Ask a group of people to raise their hands if they think suicide is more common than murder, and only a few hands will go up. Ask the same group of people to raise their hands if they personally know anyone who was murdered, and then if they personally know anyone who committed suicide. The group experience will be much more reflective of the reality.

The representativeness heuristic occurs when people overvalue the similarity of an individual event to other events, ignoring the overall likelihood of that type of event. In America, if an animal that looks like a zebra dashes by very quickly, it is much more likely that the animal was a white horse catching the shadows from the trees, no matter how similar it looked to a zebra. This is because horses are much, much more common than zebras in America. We tend to overvalue our immediate impressions and not take into account our background knowledge as much as we should.

Tversky and Kahneman have been able to make a great deal of hay out of the representativeness heuristic because there is an exact mathematical formula for how much weight to give to background experience in the face of new evidence, at least when all the probabilities are known exactly. This formula is known as *Bayes' Rule.* By giving subjects word problems where the exact probabilities for every event are included, the researchers have been able to show that the background probabilities, known as *base rates,* are systematically ignored by the vast majority of people.

Of course, most people don't know Bayes' Rule by heart, and even those who do aren't necessarily ready to do the mathematical calculations in their head every time they think they see a zebra or some other infrequent event. But these researchers have definitely shown that Bayes' Rule is not an automatic part of how our brain estimates probabilities in all cases. Given that Bayes' Rule is useful only when we have accurate statistical information about base rates, something that is rare in nature, it is not all that surprising that our brains don't automatically calculate according to Bayes' Rule.

Variant View: Undermining Heuristics and Biases

Gerd Gigerenzer and other researchers have been able to show just how fragile many of the various word problems are in detecting heuristics and biases. For example, some of these word problems generate accurate answers from subjects with slight wording changes, such as replacing the word *probability*. By asking how many of a certain event out of a hundred similar events would have the effect in question, instead of asking about the probability of the effect, Gigerenzer has prompted people to give more sensible answers.

Word problems demonstrating the *overconfidence effect*, where people overestimate the likelihood of being right when they answer a question, are also fragile. This effect disappears when people estimate how often they were right in answering a set of questions, rather than for just a single question. This makes sense, if you think about it. Suppose you have just taken a test that you found pretty easy. If you are shown one of the questions and asked, "Do you think you got this right?" you will probably say, "Yes." After all, it was an easy test, you tried your best on every question, and this particular question was no harder than the rest. So, you are probably quite confident about any question on the test. On the other hand, if you are asked, "Did you get 100 percent on the test?" you would probably say, "No." Somewhere along the line, you probably made a mistake; you just don't know where.

With an easy test, the probability of getting the right answer for any one question is very high, and, as we noted, people have trouble with very high (and very low) probabilities. And probabilities are always easier to estimate when we have a larger sample. You can't get a smaller sample than a sample of one.

There are two basic lessons to be learned from the research on heuristics and biases. The first is that people do not make very good decisions when probabilities are very high or very low. (The popularity of lotteries is a good example of this.) The second is that we overvalue our own personal experience, even our personal experience merely hearing stories about distant events.

Summary

Cognitive psychologists make extensive use of the many logical and mathematical tools developed in the last 150 years to measure our thinking in comparison to what experts say is the mathematically correct way to calculate the answers. This allows them to figure out indirectly how we think about different kinds of problems without relying upon our verbal self-reports, which can be unreliable.

Quiz

1. The method of introspection was used by the _____.

   (a) Functionalists

   (b) Logicians

   (c) Structuralists

   (d) Pragmatists

2. Which measurement technique is used by cognitive psychologists?

   (a) Reaction times

   (b) Likelihood of possibilities

   (c) Known correct answers

   (d) All of the above

3. _____ are the building blocks of thought.

   (a) Introspection

   (b) Concepts

   (c) Reaction times

   (d) Logic

4. In a categorization hierarchy, the _____ level is the default level at which we use and define objects on a daily basis.

   (a) Basic

   (b) Superordinate

   (c) Subordinate

   (d) None of the above

5. Categorization by _____ involves comparing the object to a checklist, and categorization by _____ involves comparing the object to the best example of the category.

   (a) Prototypes, features

   (b) Hierarchy, concepts

   (c) Features, prototypes

   (d) Concepts, hierarchy

6. Which stage of problem solving has the greatest impact on the speed and likelihood of finding a good solution?

    (a) Designing a solution strategy

    (b) Representing the problem

    (c) Determining when a solution is satisfactory

    (d) None of the above

7. Which of the following is a barrier to correctly representing a problem to be solved?

    (a) Irrelevant information

    (b) Unnecessary constraints

    (c) Functional fixedness

    (d) All of the above

8. A(n) _____ guarantees success if applied in the right circumstances and followed correctly.

    (a) Heuristic

    (b) Algorithm

    (c) Trial and error strategy

    (d) Rule

9. Which of the following is a theory of reasoning?

    (a) Mental logic

    (b) Mental models

    (c) Schemata

    (d) All of the above

10. The _____ heuristic is based on the idea that the more often you have heard about something, the more common it is.

    (a) Availability

    (b) Representativeness

    (c) Overconfidence

    (d) Uncertainty

PART TWO TEST

1. Contemporary psychological research on emotion tends to focus on which of the following?

    (a) Classification of emotions

    (b) Process of emotions

    (c) Communication of emotions

    (d) All of the above

2. Evolutionary theorists argue that emotions evolved for the practical purpose of _____.

    (a) Communication

    (b) Survival

    (c) Social interaction

    (d) None of the above

3. Emotional arousal can prepare us for which of the following?

    (a) Fleeing

    (b) Fighting

    (c) Feeding

    (d) All of the above

4. The modern definition of emotion

    (a) Is agreed on by the majority of psychologists

    (b) Is not agreed on by the majority of psychologists

    (c) Does not allow for scientific study

    (d) Is required for scientific study

5. According to Maslow, which of our needs is the lowest level of the hierarchy of needs?

   (a) Physiological needs

   (b) Safety needs

   (c) Self-actualization

   (d) Esteem needs

6. According to the parental investment theory, the sex that makes the _____ investment in producing and nurturing offspring tends to be less selective when choosing partners.

   (a) Smaller

   (b) Equal

   (c) Larger

   (d) Solo

7. Two areas of society in which the application of the study of motivation is most important are _____ and _____.

   (a) Education, business

   (b) Military, education

   (c) Business, military

   (d) Education, acting

8. _____ psychology addresses the points of contact between the worker and the job.

   (a) Human factors

   (b) Organizational

   (c) Personnel

   (d) Human resources

9. In basic classical conditioning experiments, salivation to the bell is the _____.

   (a) Unconditioned stimulus

   (b) Unconditioned response

   (c) Conditioned stimulus

   (d) Conditioned response

10. _____ is the most efficient way to establish an operantly conditioned response.

   (a) Forward conditioning

   (b) Backward conditioning

   (c) Shaping

   (d) Random conditioning

11. Grounding a teenager for earning bad grades (to discourage this behavior) is an example of what operant conditioning strategy?

   (a) Negative punishment

   (b) Positive punishment

   (c) Negative reinforcement

   (d) Positive reinforcement

12. Because connectionist models are designed to resemble groups of interconnected nerve cells, they are often called _____.

   (a) Symbolic models

   (b) Neural networks

   (c) Cognitive models

   (d) All of the above

13. _____ is another term for a child's first sentences.

   (a) Telegraphic speech

   (b) Over-regularization

   (c) Cooing

   (d) Holophrases

14. The _____ of language involves using the structure of language appropriately.

   (a) Pragmatics

   (b) Semantics

   (c) Grammar

   (d) Meaning

15. The evidence that children across cultures learn language very similarly supports which theory of language acquisition?

   (a) Social learning

   (b) Learning

(c) Integrative

(d) Nativist

16. The key to the rapid growth in vocabulary between one to six years of age involves _____.

    (a) Holophrases

    (b) Fast mapping

    (c) Telegraphic speech

    (d) Overregularization

17. Short-term memory can hold an average of _____ chunks of information at any given time.

    (a) Five

    (b) Seven

    (c) Nine

    (d) Infinite

18. Which type of memory has the largest capacity?

    (a) Sensory memory

    (b) Short-term memory

    (c) Long-term memory

    (d) Episodic memory

19. _____ memory is the type of memory that is deliberately brought to mind.

    (a) Episodic

    (b) Implicit

    (c) Semantic

    (d) Procedural

20. Automatic encoding is _____.

    (a) Learned

    (b) Automatic

    (c) Both a and b

    (d) None of the above

21. In a categorization hierarchy, the _____ level is the least inclusive level.

    (a) Basic

    (b) Superordinate

    (c) Subordinate

    (d) None of the above

22. Categorization by _____ involves comparing the object to the best example of the category.

    (a) Prototypes

    (b) Hierarchy

    (c) Features

    (d) Concepts

23. The _____ strategy in designing a solution involves comparing your current position with a desired endpoint and then trying to find a means of closing the gap between the two.

    (a) Sub-goal

    (b) Means-end

    (c) Backward search

    (d) Functional fixedness

24. If you don't know the probabilities of all potential outcomes, you must use the theories of decisions under _____.

    (a) Risk

    (b) Heuristics

    (c) Representativeness

    (d) Uncertainty

25. In a categorization hierarchy, the _____ level is the most inclusive level.

    (a) Basic

    (b) Superordinate

    (c) Subordinate

    (d) None of the above

PART THREE

From the Outside In

The answers to many of psychology's most profound questions lie in an understanding of the physical processes that make them possible. Psychology, after all, studies what is unique to animals. It is the nervous system of the animal, along with affiliated bodily systems, that enable animals to cope with the world in ways that plants and inanimate objects cannot. Part Three talks about what psychology has discovered about how the body makes psychological processes possible.

We have ordered these chapters from the outside of the body inward to the brain. Chapter 15, "The Study of Sensation," explores what psychologists have discovered about how information about the world is delivered to the nervous system. Chapter 16, "The Study of Perception," talks about how the nervous system integrates that information in order to provide us with a coherent picture of the world. Chapter 17, "Neuroscience and Genetics," covers what has been discovered about the nervous system and our DNA and how all of what has been covered in the previous chapters is made possible.

CHAPTER 15

The Study of Sensation

In order to cope with the world, we need to know what is going on in the world. Our five traditional senses—sight, hearing, touch, taste, and smell—plus our internal senses, such as balance, movement, and position, are the means by which information about the world reaches our brains and makes behavior possible.

Overview

Only certain information about the world is available to our nervous system. So far as we know, all of this information enters our systems through the senses. At the very periphery of our nervous systems are special cells, called *transducers*. Instead of receiving signals from other nerve cells, these cells become active when specific changes happen in their environment. Then they send signals to other nerve cells, passing along information about those environmental changes further into the nervous system.

Different types of transducers are sensitive to different types of environmental changes. Some are sensitive to light, some to vibration, some to chemicals, some to physical contact, some to temperature, and so on. Depending on where these transducers are located in the body and how they are located within the sense organs, they pick up specific types of important bits of information and relay them to the brain. The sense organs are specialized organs, containing transducer cells, structured so as to convert the physical energy from important environmental events into neural signals via the transducer cells.

Psychologists are interested in how the brain makes sense of physical stimuli and how what we perceive may be different from the sensory input we receive. The first step in this study is to understand sensation. The next step (covered in the next chapter) is to understand perception.

HISTORY

The study of sensation is one of the oldest areas of psychology. In part, this is because the sense organs are at least partially at the exterior of the body. Long before psychological inquiry became scientific, people had figured out that the eyes conveyed vision, the ears conveyed sound, and so on. (Remember peek-a-boo?) Aristotle listed the five classic senses: sight, hearing, taste, smell, and touch. Current science has subdivided the sense of touch and added a few other related senses.

The first scientists to investigate sensation were not psychologists, but anatomists and physiologists. Because most of the sense organs had been identified in prehistory, the obvious first step in investigating sensation was to examine the sense organs. The microscope was a key development in this regard. The nineteenth century discovery, for instance, that the retina was composed of separate, tiny light receptors, began an inquiry into vision that is still underway. After all, our subjective impression of the visible world is in terms of continuous expanses of color and light, not little pinpricks of color and light. Even the invention of television and color printers relied upon the discovery that, if the pinpricks (or pixels) are small enough and close enough together, humans perceive the result as a picture.

After the physiologists came the first psychologists interested in sensation, the psychophysicists. Psychophysics is one of the oldest areas of psychology, and it is still around. Psychophysicists try to find mathematical relations between the physical regularities of the world and their effects on the senses.

Variant View: The Inefficiency of ESP

An interesting exception to the two-pronged strategy in the study of sensation is the case of psychic phenomena, or ESP. ESP stands for *extra-sensory perception* because an early theory was that people were able to obtain knowledge of the world without using any of the senses. Some more recent views argue that psychic phenomena might operate by use of thus far undiscovered sense organs. Because the means by which the information is conveyed to the nervous system is unknown, research into psychic phenomena has been restricted to the second strategy, that of comparing the patterns of environmental events to the reports of sensations presumably produced by those events.

Even if we assume that psychic phenomena exist, the many decades of psychic research have produced a very curious result. Consistent relations between the environmental events and the psychological sensations are extremely hard to detect and measure. Outside the laboratory, it is very unusual, indeed, extraordinary, for people to obtain knowledge of the world without use of the known senses. Inside the laboratory, the amount of information conveyed appears to be tiny at best, at least when compared to any of the other senses. For instance, touch is one of the least informative of the senses. Using touch, however, we can easily determine the results of an entire series of coin flips, heads or tails, with very few errors. This is something far beyond the reach of even the most powerful psychics. In other words, psychic powers, if they exist, are an extremely inefficient means of obtaining information.

An evolutionary advantage of psychic phenomena, if they exist, would be the ability to obtain information found beyond the reach of the ordinary senses. It is unclear why this evolutionary advantage would fail to produce highly efficient information delivery over time. If psychic abilities exist, they are relatively young in evolutionary terms, or else there are extreme limitations to how much information can be conveyed without the senses.

CURRENT APPROACHES

Psychologists usually take a two-pronged approach to studying sensation. After the sense organ is identified, they investigate the mechanics of how physical energies are converted into neural signals. In addition, psychophysicists investigate the mathematical relations between the character of the physical energies and the sensations produced. How do different environmental events appear differently to us?

Psychophysics

The outside world is made up of matter and energy (as are we). The senses obtain information about the outside world by being sensitive to different types of energy and, in particular, changes in energy. When that information arrives in the brain, we are often able to report on it. The ability to say something about what we sense is a psychological matter. The neurological signals and psychological information created from the physical energies make up our picture (or representation) of our environment. Like any picture, it is not exact. There are many things about our environment we do not and cannot know. How our psychological picture of the world relates to the physical world is the subject of psychophysics.

Psychophysicists have developed careful methods for discovering the relationship between what is actually out there and what people experience. They use mathematical models to describe that relationship. The most basic relationship between physical events and psychological reports is that of detection. When there is light, do we see it? When there is sound, do we hear it? Determining the circumstances under which a sensory signal is detected has been the subject of a series of mathematical theories.

THRESHOLDS

The first theory used by psychophysicists was the *absolute threshold*. What is the smallest amount of light or sound or flavor, etc., that makes for a detectible image or noise or taste? This amount is called the threshold. This threshold varies

Key Point: Measuring Absolute Threshold

Experiments for measuring the absolute threshold should be simple. You start with a very dim light, or a very quiet sound, or a very soft touch, etc., and then you gradually increase it until the subject reports sensing it. As it turns out, things are not that simple. There is no point where the subject goes from never detecting the signal to always detecting it. Instead, over a range of stimulus intensities, the subject gradually detects the signal a larger and larger percentage of the time. There is a point where the signal is so weak that the subject never detects it and a point where it is so strong that the subject always detects it, but in between, there is a range where the subject detects the signal only some of the time. Within that middle range, the point where the subject detects the stimulus 50 percent of the time is defined as the absolute threshold.

Why 50 percent? Why not 25 percent? Or 80 percent? The bottom line is that there is no reason. Fifty percent is an arbitrary amount. The only justification is that, in a way, 50 percent is the halfway point between never detecting the stimulus and always detecting it.

depending upon the color of the light, the frequency of the sound, the nature of the flavor because the person is distinguishing between the background noise and the stimulus. The stimulus must have enough of its defining quality (brightness of light, intensity of sound, intensity of flavor, etc.) so a person notices its presence.

The next notion was the *just noticeable difference* or *jnd* (also called a *difference threshold*). This notion was based on the observation that a tiny pinpoint of light in a darkened room is more noticeable than that same pinpoint outside on a sunny day. Likewise, we may be able to hear a pin drop in a perfectly quiet room, but not during a performance of a Beethoven symphony or during an artillery attack. The amount of change in physical energy that is noticeable, the *jnd*, depends upon the total amount of energy present.

Sometimes, a person does not simply need to detect the presence of a stimulus, but rather to differentiate among stimuli of different magnitudes. Consider a pin dropping in a room where other pins are also dropping. If ten pins are dropped all at once, could you hear an eleventh pin at the same time? What if there were 100 pins?

The jnd is the size of the difference in a quantity of the stimulus needed for the person to notice a difference between the two stimuli. Thus, when evaluating the brightness levels in two lights, the just noticeable difference is the minimum difference needed to distinguish between the two. If you cannot detect one or two additional pins when there are ten pins dropping, but can detect the difference between ten pins dropping and thirteen pins dropping, then the jnd for ten pins is three.

Weber's Law states that a constant percentage of a change in the magnitude of a stimulus is necessary to detect a difference. So, if the jnd for ten pins is three additional pins, then the jnd for 100 pins should be 30 and for 5,000 pins should be 1,500. The next largest bunch of dropping pins that will sound different will be a bunch with 30 percent more pins. Weber's Law, like other early psychophysical laws, is remarkably accurate up to a certain point, but there are exceptions where it does not hold.

SIGNAL DETECTION

Signal detection theory allows the estimation of sensory sensitivity without determining the cause in advance. After the sensory sensitivity is measured, the degree

Big Background

Psychophysics advanced from absolute thresholds, where sensitivity differs with sensory qualities, to just noticeable differences, where sensitivity also differs with total sensory quantity. The next advance was a general approach for determining what other things might affect sensory sensitivity.

Variant View: A False Sense of Precision

The advantage to signal detection theory is that it does not assume any particular cause in advance. The problem is that many of the things that can affect sensory sensitivity, including the severity of the consequences of error, how tired the subject is, how long the subject has been observing, the subject's mood, even the subject's health and well-being, are all things that are harder to measure than the physical energies that make for sensation.

These mathematical models used in psychophysics are, in a way, almost too powerful. The relationship between the physical and the psychological can be measured more precisely than the psychological causes that change it.

of sensitivity can be compared to various possible causes to see which one is the actual cause.

Signal detection theory seeks to explain why people detect signals in some situations but not in others. Signals are always embedded in noise. This noise is an event or other stimuli that can be mistaken for the stimulus. The challenge is to distinguish the signal from the noise. The interesting psychological phenomenon with signal detection is that we are not always accurate. First, one must consider the threshold level for distinguishing between a stimulus and the background noise. The lower the threshold, the greater the sensitivity. Second, a person's bias, or her willingness to report the presence of a stimulus, can come into play. People change their bias by adjusting their concept of how strong a signal needs to be before they will say that they detect it.

Both sensitivity and bias can be assessed by comparing the occasions in which a person reports the presence or absence of a stimulus with the occasions when the stimulus is actually present or absent. Thus, it is a comparison between signal detection and reality. Table 15-1 illustrates the four possibilities. The beauty of signal detection theory is that the concepts can be applied to a variety of topics in which there may be a difference between reported perceptions and reality.

Table 15-1 Possible Outcomes of Signal Detection

		Is the signal really there?	
		Yes	**No**
Is the signal reported as being perceived?	**Yes**	Hit	False alarm
	No	Miss	Correct rejection

Vision

Vision is the richest of the human senses. Our eyes receive light from the surrounding objects and translate it into nerve impulses that travel to the brain. When they get there, they are organized and you experience shapes, colors, textures, and movement. In this section, we will discuss the details of the visual system in terms of the nature of light and the structure of the visual system. Finally, we will discuss three theories of color vision and how psychologists reconcile these three theories.

Tips on Terms: Waves

Both light and sound are composed of waves. Waves are regularly fluctuating streams of energy that travel through space. Figure 15-1(a) shows a simple wave. At each point, the amount of energy changes over time in a regular, repeating pattern called a *cycle*. The number of cycles per second is the *frequency*. Waves tend to travel at a fixed, steady speed, so a specific frequency means that the cycles are a fixed distance apart. The distance between the start of the cycle and its end is the *wavelength*. Thus, the wavelength is a measure of the frequency. The shorter the wavelength, the higher the frequency; the longer the wavelength, the lower the frequency. The difference between the maximum and minimum amounts of energy (the height of the wave) is a measure of the *amplitude*.

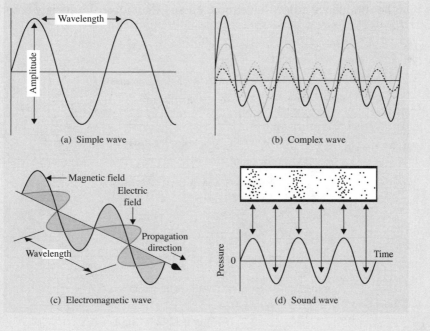

(a) Simple wave   (b) Complex wave

(c) Electromagnetic wave   (d) Sound wave

Figure 15-1   Waves

continued...

Waves are measured with wave functions, which are graphed as energy (or a related value) over time. The sine curve is the basic wave function, representing one constant frequency at one amplitude. All other waves, no matter how complex, can be represented as combinations of sine waves. Figure 15-1(b) shows a complex wave made up of three sine waves.

Light is a form of electromagnetic wave shown in Figure 15-1(c). The energy is in the form of changing electrical and magnetic charges that travel in packets called *photons*. Photons can travel through a vacuum, so light does not require a medium. Sound is a pressure wave. The energy is in the form of changing compression of a medium, such as air or water, through which the sound travels. Figure 15-1(d) illustrates a sound wave.

THE NATURE OF LIGHT

Light is made up of subatomic particles of light radiation (called *photons*) that travel in wave-like patterns. As shown in Figure 15-2, different colors of light have different frequencies and wavelengths. With higher frequency, the peaks of the wave arrive more often, and the length of time between peaks (the wavelength) is shorter. Violet has the highest frequency. Red has the lowest. A light's brightness is determined by the intensity, or the *amplitude*, of the light wave. This is defined by the height of the wave, which measures the amount of energy. At the same intensity, higher frequency colors have more energy than lower frequency colors.

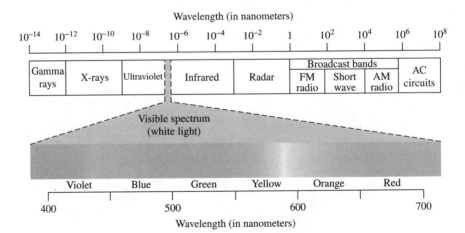

Figure 15-2   Light

Fun Facts: Visible Light

In terms of physics, visible light is electromagnetic radiation whose frequency falls within a very narrow range (380–760nm). Cosmic rays, X-rays, and ultraviolet radiation have higher frequencies and are invisible to the eye. Infrared, heat, and microwave radiation have lower frequencies and are also invisible. While some other animals can see in slightly higher and/or lower frequencies than we can, the vast majority of frequencies are outside the visible range of all earthly animals.

In comic books, and some psychology texts, we speculate as to how wonderful it would be to be able to see in these vast frequency ranges. But, in fact, there is an important evolutionary advantage to being able to detect electromagnetic radiation in the very narrow range we call light. The radiation in this range, and only in this range, does not pass easily through solid objects. Instead, it is absorbed by them or reflected off of them. Seeing visible light has the unique advantage of helping us detect physical objects.

THE STRUCTURE OF THE EYE

The visual system begins with the eye, through which light waves enter. The following list highlights the major structures of the visual system, with a particular focus on the structure of the eye:

- **Cornea**   A transparent rigid structure that is curved over the front of the eyeball. It takes the incoming light wave and bends it in order to bring it into focus.

- **Iris**   A ring of pigmented tissue that gives the eye its color. Contraction and relaxation open and close the pupil.

- **Pupil**   The black opening in the center of the eye. By opening and closing the pupil, the amount of light entering the visual system is regulated.

- **Lens**   A transparent elastic structure located behind the pupil. It allows the eye to adjust its focus in accordance with the object's distance.

- **Retina**   The inner surface at the back of the eyeball that is lined with visual receptors.

- **Fovea**   The central part of the retina, which contains densely packed cells that transform light into nerve impulses. This area has the highest density of cones and allows for the highest resolution of vision.

- **Rods**   Specially adapted receptor cells that are used for night vision and dim-light vision. A single photon of light can stimulate a rod to fire. Each eye contains between 100 million and 120 million rods, and they are present everywhere except the fovea.

- **Cones**   Specially adapted receptor cells for bright light and color vision. It takes many more photons to stimulate cones to fire than it does for rods to fire. Each eye contains between 5 million and 6 million cones, and they are densest near the fovea.

- **Optic nerve**   The axons from retinal nerve cells in each eye are gathered into a single large cord. This cord carries impulses from the retina into the brain.

THEORIES OF COLOR VISION

The colors we see are not inherent in the objects we see. The sky is not blue. The colors we see are essentially products of how our visual system interprets the different wavelengths of light that reflect off the object and hit the eyes. If the wavelength is 670 nm, you see red. If the wavelength is 530 nm, you see green. Wavelengths are a continuous span of values. The boundaries between colors in the color spectrum are not found in light, but only in the eye. In terms of wavelengths, the difference between a deep red and an orange-red is less than the difference between yellow and blue.

With the exceptions of lasers and a few other specialized forms of light, every color we see, either natural or human-made, is a mixture of light of different wavelengths. Depending on the mix, we see different colors. The sky is not blue; our sensation of the sky is blue.

Key Point: Adding and Subtracting Colors

As children, we all learned how to mix paint to make new colors. We learned early on that blue and yellow make green and that yellow and red make orange. This is true when describing pigment mixing. Here, however, we are describing how light is mixed, and light works in almost the opposite way from pigments. Pigments use the process of *subtractive mixing*. Pigments, such as in paint, absorb some wavelengths and selectively reflect back the rest, which is what we see. When we combine pigments, more types of light get absorbed and a smaller remainder is reflected. Light, on the other hand, uses the process of *additive mixing*. This involves adding, or superimposing, the wavelengths of light. When we combine differently colored lights, each light adds something to the color we see.

The theories of color vision explore how the visual system translates different wavelengths of light into the perception of color. Here, we will describe three theories and then discuss how to reconcile these apparently disparate theories.

Trichromatic Theory

The *trichromatic* theory (or the *Young-Helmholtz* theory) of color vision holds that there cannot be a separate receptor for each shade of each color that we perceive. Instead, these theorists argued that there must be three types of cones, each receptive to a certain frequency of light. Our "blue" cones are sensitive to short-wave light; our "green" cones are sensitive to medium-wave light; and our "red" cones are sensitive to long-wave light. By mixing just three primary colors of light, it is possible to make every color in the visible light spectrum. Color mixing is based on the ratio of the types of cones that are activated at a given time. For example, when the red and green cones are equally more active than the blue cones, we perceive yellow. When all three types of cones are equally active, we perceive white.

The problem with the trichromatic theory is that it cannot explain *afterimages.* Afterimages are a visual impression that persists after the visual stimulus is removed. In a classic example, a person stares at a version of the American flag in which green stripes replace the red stripes, black stripes and stars replace the white stripes and stars, and a yellow field replaces the blue field. After staring at the flag for a few moments, the person is exposed to a white screen. The person reports seeing a faint image of a properly colored American flag superimposed on the white screen. According to the trichromatic theory of color vision, because the white screen is stimulating all the cones equally, the person should see no color. Thus, the trichromatic theory of color vision cannot explain this phenomenon.

Opponent-process Theory

The *opponent-process* theory of color vision proposes that there are four primary types of cones. In addition to the red-sensitive, green-sensitive and blue-sensitive cones, the opponent process theory claims that there are also cones that are sensitive to yellow light. This theory also argues that these four cone types are organized into two opponent systems, a red-green system and a blue-yellow system. The red-green system contains cells that are excited by red light and inhibited by green light, and cells that are excited by green light and inhibited by red light. The blue-yellow system contains cells that are excited by blue light and inhibited by yellow light, and cells that are excited by yellow light and inhibited by blue light. Like the trichromatic theory on which it builds, all colors in the visual spectrum can be accounted for by the relative ratio of activated cones.

One advantage of the opponent-process theory over the trichromatic theory is that it can explain afterimages. When a person looks at the green stripes, the green-sensitive cones in the red-green opponent system are stimulated. Soon, this response adapts to the stimulation and becomes less vigorous in that it takes more stimulation to make the cells fire. When the person looks at the white surface, it normally looks white because the light reflected from it stimulates the red and green cells equally. Because the green response in the striped pattern has been suppressed through adaptation, the red response dominates and he sees a red stripe instead. The same is true for the blue-yellow system. When a person stares at the yellow field, he saturates the cones stimulated by yellow light. When exposed to the white surface, the blue cones require much less light to stimulate the cells to fire, and the person sees a blue field instead.

A problem with the opponent-process theory is that it cannot explain *color constancy*. Color constancy is the tendency for an object to appear to be nearly the same color under a variety of light or lighting conditions. This generally applies to familiar objects. For example, when you park your red car in the daylight and return to it at night, you do not perceive it as a black car even though that is the information your visual system is receiving. Conversely, an unfamiliar object, such as a new sweater, appears to change color when you look at it in natural light versus indoor lighting. The constancy of color for familiar objects cannot be accounted for by the opponent-process theory.

Retinex Theory

The *retinex* theory argues that we perceive color through the combined contributions of the retina and the cortex (hence the theory's name). The cortex compares patterns of light from the retina and puts together a color perception for each area. However, because the cortex has experience perceiving the color of familiar objects, it expects something to be a certain color. Therefore, with the help of the cortex, this theory can explain color constancy.

Reconciling the Three Theories

Given that these three theories of color vision explain various aspects of color vision, which theory is correct? Actually, each of them is correct. Just because there is more than one theory for a process does not mean that only one theory is correct and the others are wrong. In the case of color vision, each of these theories examines the phenomenon at a different level of explanation. The trichromatic theory explains color vision at the rod and cone level. The opponent-process theory explains color vision at the bipolar cell and ganglion cell level (nerve cells found in

Big Background: Competing Theories

Several times in *Psychology Demystified,* we have seen where two theories thought to be alternative views of some phenomena both turn out to be true. The notion that science advances when a new theory comes into conflict with an old one dates back to the nineteenth century, but it really took hold in the twentieth century. Not only did it impact philosophers and historians of science, but it eventually came to alter the way that science and other academic endeavors were actually practiced. When additional information, such as afterimages, could not be accounted for by the current theory, a new theory was not presented as an addition to the current theory, but as a competitor to it. No effort was made to see if the theories actually contradicted one another. Instead, it was assumed that both could not be true. For almost a hundred years, psychologists on both sides of the debate assumed that only one theory could be right, simply because some philosophers said that was how truly "scientific" theories operated. Happily, in the case of the trichromatic and opponent-process theories, we actually have two separate types of brain cells that have been found and shown to implement the two theories at the same time.

The important lesson from the case of the three theories of color vision is that both the brain and the psychological phenomena it produces are terribly complex. Traditional psychological theories tend to explain only a few phenomena. Different parts of the brain are likely to be doing different things, implementing different parts of different theories. Philosophers have known for a long time that not all theories compete. Psychologists need to understand that as well.

the retina behind the receptor cells and in the thalamus). Finally, the retinex theory explains color vision at the cortical level. Therefore, each theory is correct, and the process of color vision is more complex than we originally thought.

Hearing

The sense of hearing detects sound waves just as the sense of sight detects light waves. Beyond that, there are important differences. Vision operates in a straight line. The focusing apparatus of the human eye means that we see what we point our eyes at. The ears, on the other hand, are nondirectional. This combination of sense provides us with important evolutionary advantages. We can see farther than we

can hear, but only if nothing is obstructing our vision. We can hear things that we are not facing. We can hear around corners and over hills. A sound can warn us that there is something worth looking at. Because vision provides much more information (in terms of the amount of neural signals) than hearing, we can use vision to confirm what we hear.

THE NATURE OF SOUND

Sound waves are different from light waves. Sound waves are changes in pressure. In order for sound waves to exist, there must be some substance to be under pressure. Most often, for humans, this substance is the air around us. (We can also hear underwater and through walls, etc.) The substance that carries the sound is called the *medium*. In part, this is why sounds travel differently than light. Light always travels in a straight line, even when it bounces off something. Sound not only bounces; it also spreads out through the medium. The pressure changes in the air around the corner affect the air on our side as well. We don't hear as clearly around corners, but the sound flows out to us through the air.

The same mathematical function, the sine wave, which measures the electrical and magnetic charges that make up light waves, can also be used to measure the changes in pressure that make up sound waves. Sounds with larger amplitude waves tend to be louder. Sounds with shorter wavelengths tend to be higher pitched. However, the match between amplitude and loudness and between frequency (wavelength) and pitch is not exact. For example, if two sounds have the same amplitude, the higher frequency sound will tend to sound louder to us. Just as pure frequencies of light are uncommon, so are pure tones. Most sounds are complex waves.

Fun Facts: What We Can Hear

Just as we cannot see electromagnetic radiation when the frequency is too high or too low, we cannot hear vibrations when the frequency is too high (above 20,000 vibrations per second) or too low (below 20 vibrations per second). As with light, other animals can hear sounds with higher and lower frequencies than we can. As with light, the range of frequencies we can hear are very useful ones. The frequencies we can hear surround the frequencies of the sounds we make when we speak.

Intriguingly enough, there are other aspects of sound that can make them hard to detect. If a sound is very brief and changes frequency very rapidly, it requires specialized brain cells to detect it. Many of the consonants of human language (those sounds other than *a, e, i, o,* and *u*) have exactly this property, and humans have the specialized brain cells to detect them.

While the eyes have evolved to help us cope with the physical world, the ears seem to have evolved to help us cope with the social world.

THE STRUCTURE OF THE EAR

The auditory system begins with the ear. Sound enters the ear and is transduced into nerve signals using a series of mechanisms within the ear. The following list highlights the major structures of the auditory system:

- **Pinna (also called the auricle)**   The visible part of the outer ear. Sound is collected there and directed into the outer ear canal.

- **Outer ear canal**   A tube through which sound travels to the eardrum.

- **Eardrum (also called the tympanic membrane)**   A thin membrane, stretched across the end of the ear canal, which divides the outer ear from the middle ear and vibrates due to the pressure of the sound waves.

- **Hammer, anvil, and stirrup**   Three tiny bones, shaped like their names, that pass the vibrations through the middle ear from the eardrum to the cochlea.

- **Cochlea**   A spiral-shaped tube in the inner ear that doubles back on itself. It is filled with fluid and lined with cilia (tiny hairs) that move when membranes above and below them vibrate due to pressure changes in the fluid.

- **Hair cells**   The transducer cells tipped with cilia. Movement of the cilia causes the nerve cells to send signals to the auditory nerve. The auditory nerve passes these signals to the brain.

THEORIES OF PITCH

Just as there is more than one theory about how the frequency of light creates the sensation of color, there are two theories about how the frequency of sound creates the sensation of pitch. Just as with vision, each theory accounts for some, but not all, of the phenomena, and just as with vision, twentieth century theorists insisted on viewing the theories as incompatible alternatives. However, we now think that both mechanisms work together using different neural mechanisms.

The first theory of pitch, *place theory*, is based on the notion that different parts of the membranes in the cochlea vibrate when different frequencies pass through the fluid. (This is due to the shape of the cochlea.) Only the cilia located near the part of the membrane that is vibrating send signals to the brain. The problem is that, at very low frequencies, the entire membrane vibrates.

The second theory of pitch, *frequency theory*, is based on the notion that each time the cilia move, they create a nerve signal. The higher the frequency, the faster the membranes vibrate; the faster the membranes vibrate, the more signals the cilia send. The problem is that nerve cells cannot signal as fast as sound vibrations can happen.

Currently, theorists believe that frequency theory is true for low-pitched sound, place theory is true for high-pitched sound, and both are true for middle-ranged sounds.

Location-based Senses

We see only through our eyes and we hear mostly through our ears. However, the location of our eyes and ears on our bodies is irrelevant to what we see and hear. We have a number of other senses where what we sense is very relevant to where we sense it. Some of these senses are located in the skin and tell us about external objects that are touching various parts of our bodies. Other senses are located internally and tell us about our body's posture and its position relative to the world. The senses that relate to location are called the *somasthetic* senses.

WITHIN THE SKIN

The senses based on transducers in the skin, called the *cutaneous* senses, tell us about the relation between objects and our bodies. Because our skin surrounds us on all sides, these senses convey information about the location of objects along with whatever else they tell us.

The basic categories of transducers in the skin are those that support the detection of pressure (touch), temperature, and pain. While there are clearly different types of receptors, the exact relationship between the different receptors and the different sensations is currently unclear. Different types of signals, in combination, from different types of receptors give rise to the many different types of touch and pain and temperature sensations.

In addition, even more so than with the other senses, the dividing line between sensation and perception for these skin senses is unclear. Some of these sensations are more or less directly transduced, and others are clearly calculated further along in the nervous system. In the case of vision, we can talk categorically about extracting

---

Tips on Terms: Cutaneous Senses

The senses of touch, temperature, and pain are called the *cutaneous* senses because the bulk of the receptors is in the skin itself. Some of these receptors, however, are inside the various organs and muscles. That is how we are able to feel pain and pressure inside our bodies. We have far more of these receptors in the skin than elsewhere.

Tips on Terms: Texture

Texture is a quality we detect by touch. Differentiating textures is almost certainly a question of perception. Our brains combine different kinds of signals over time to determine the texture of what we are feeling. We know this because we usually move our skin across the textured object in order to determine the texture. Motion creates this pattern of changing signals of various types over time.

objects from the background and then identifying what they are. We can talk about audible sounds in the same way we talk about visible objects. The notions of object and sound help make the distinction between sensation and perception clearer. Our skin helps us identify objects, injuries, conditions, and a host of other things, however, so the situation is much more complex.

Touch

Touch is the only one of the somasthetic senses identified by Aristotle. Touch operates because of pressure sensitive receptor cells in the skin. Some of these receptors detect light pressure and some detect deep pressure. Some appear to detect different types of movement of the skin such as sideways displacement and stretching. Some receptors detect rapid movement, such as that caused by vibrations.

Temperature

It is obviously very important to be able to detect temperature. Extremes in temperature can be harmful, even deadly. What we need to do to protect ourselves from heat and cold differ. There is one type of temperature receptor that signals when it detects heat. There is another receptor that detects extremes in temperature, either

Extra Exercise: A Trick of the Temperature

Because there are two different kinds of temperature receptors, we can trick them. If we touch both cold and comfortably warm objects in close proximity, we sense extreme heat where there is none. This is like a tactile illusion, analogous to visual illusions. One way to feel this effect is to step into a warm shower or bath after coming in from outdoors on a very cold day when your hands and feet are still cold. The warm water will feel very hot on the cold parts of your body, and warm elsewhere.

hot or cold. The two of these combined are enough to tell us all we need to know: Heat without extreme equals warm. Heat with extreme equals hot. No heat without extreme equals cool. No heat with extreme equals cold.

Pain

There are some skin receptors that are critically involved in pain, but unlike pressure and temperature receptors, they do not transduce pain directly. Environmental pressure and temperature activate pressure and temperature receptors, causing us to feel pressure and temperature. But pain is not an environmental event. Pain receptors detect various conditions, such as extremes of other sensations (light, sound, pressure, temperature, and so on) and damage to tissues, and send signals about those conditions further into the nervous system. It is there that the neural activity that causes us to feel pain occurs. In other words, pain is always a matter of perception, not just sensation.

Like any other perception, pain is more readily affected by other conditions, including our state of mind, than are the true sensations. This is why things such as meditation and distraction can lessen pain and why anxiety and expectations can increase it.

Finally, just as with emotion, there are two pathways for transmitting pain signals into the brain, one fast and one slow. This makes sense, because pain can signal danger, so the quicker, the better. On the other hand, when there is no immediate danger, it is very important to be able to determine the type of pain (sharp, dull, stretching, strain, burning, and so on), and that takes a little more time.

INSIDE THE BODY

There are at least two more senses inside us. One of these, the *vestibular* sense, is controlled by a specialized organ, just as vision, audition, smell, and taste are controlled by the eyes, ears, nose, and mouth. This organ exists deep within the inner ear. The other, the *kinesthetic* sense, is due to receptors spread throughout the body, similarly to those for temperature, pressure, and pain.

Fun Facts: Why Scratching Works

There are specific sensations, such as itching, that are due to activity of skin receptors in some unknown combinations. These may include pain and pressure and perhaps even temperature. Whatever the combination, we know that pain blocks itching because that is how scratching works.

The thing these two senses have in common is that, together, they tell us about where our body is, both in relation to itself and to the world.

Vestibular Sense

The vestibular sense locates us with respect to our environment. It does so using the effects of gravity and inertia. There are actually two connected organs in the inner ear controlling the vestibular sense. There are the *semicircular canals,* three looping tubes that terminate in three bulbs, called the *vestibular sacs.* Like the cochlea, they are cavities filled with fluid and lined with hair cells sensitive to the fluid's motion.

The vestibular sacs determine the position of the head, because the hair cells are differentially sensitive to the pressure of the fluid due to gravity. We can tell whether our head is upright, on its side, or upside down, due to this sense.

The three semi-circular canals each loop along different directions—top to bottom, side to side, and back to front. When the head moves, the fluid inside the canals moves more slowly than the head, because liquids move more slowly than solids. (This is why drinks spill from upright cups *after* the cup stops moving.) Depending on the direction the head moves, the fluid in each of the three canals moves more or less. As the fluid sloshes around, it moves the hair cells, which detect movement of the head.

Kinesthesis

The kinesthetic sense is controlled by receptors in the joints, which detect how much each joint is bent, and in the muscles and tendons, which detect how much each muscle is expanded or contracted by sensing the tension. This allows us to sense the position and motion of each part of the body relative to the others.

The vestibular sense detects the position and motion of the head relative to the world. The kinesthetic sense detects the position and motion of each part of the body (including the head) to the others. Using them together, we are able to detect the position and motion of each part of the body relative to the world.

Fun Facts: Dizziness

When we spin around for long enough, and then stop, the fluid in the semicircular canals takes a while to settle down. This is what we call *dizziness.* Our eyes tell us we are lying still, but our vestibular sense tells us we are still spinning. Dizziness is another sort of illusion.

Big Background: Are There Other Senses?

Some animals have senses that humans do not have. Some detect heat. Some detect vibrations outside the range of sound. Some detect echoes of high frequency sounds they produce. Some may detect gravitational changes. Some detect energy patterns we do not understand well at present.

Birds and sharks appear to be able to detect magnetism. There does not appear to be a separate sense organ for magnetism. Instead, there appear to be specialized brain cells that contain tiny magnetized bits of metal. These brain cells act as transducers, conveying magnetic information directly to the brain. Intriguingly, humans may also be able to detect magnetism, as these magnetic brain cells have been found in humans.

The Chemical Senses

The remaining two of the traditional five senses, taste and smell, work together to detect different chemicals. They are called the *chemical senses.*

Chemicals provide important clues about our environment. There are dangerous gasses, of course. We may be able to smell smoke before we see fire. It is important that we do not ingest rotten food or poisonous materials. Beyond protecting us from danger, taste and smell can help us ensure that we are getting sufficient nutrients in our food and drink. Chemicals produced by people called *pheromones* are important in sex.

TASTE

Taste resembles vision, hearing, and olfaction (smell) in that it works with a localized organ, the tongue. Taste resembles touch in that the source of the sensation must come into direct contact with the organ. We only taste what comes into our mouths. Even then, the mouth must be moist enough to dissolve enough of the material so that it can react with the taste receptors. The different types of taste receptors detect salt, sweet, sour, bitter, and a recently discovered sense, called *unami,* that corresponds to meatiness. Apparently, most taste receptors are not uniquely linked to just one taste. Taste receptors seem to detect several basic flavors, but are most sensitive to just one or another.

Fun Facts: Rumor as Science

For decades, high school and college students were taught that there are separate regions on the tongue for sweet, sour, bitter, and salt. There were even in-class experiments to demonstrate this. The experiments always fail, but the practice and the theory go on. In fact, the tongue is *not* divided into specialized sections. All different types of flavor receptors appear at all places on the tongue.

Where did this error come from? In 1942, the famous Harvard psychologist and predominant historian of psychology throughout the twentieth century, Edwin G. Boring, apparently mistranslated a German article on taste written by a psychologist named Hanig. No one ever proposed that the tongue had specialized regions, much less demonstrated it empirically. Boring did not have the translation checked, and for the remainder of the twentieth century, no one else checked either. Your textbook may still contain this error.

Salt and sweet and unami are important for detecting nutritious foods. Sour may help in detecting spoiled foods that have become more acid. (Vitamin C, an important nutrient, is an acid and also tastes sour.) These receptors are based on sensitivity to specific chemicals. The value or dangers of these types of chemicals have remained relatively constant over evolutionary time. Bitterness is another matter. Many toxic chemicals found in nature taste bitter, despite the fact that these chemicals have nothing in their chemical structure in common. This means that there must be many different kinds of chemical sensors that all produce a bitter sensation. Presumably, as various plants and animals evolved to produce toxic chemicals to protect themselves from being eaten, animals with taste buds coevolved different sensors to detect those novel toxins. But all of them taste bitter.

SMELL

The sense of smell (*olfaction*), like vision and hearing, detects things at a distance. Instead of waves, it detects gasses diffused in the air. When we smell a solid or a liquid, it is the slight gaseous emanations from the surface that we actually smell. Smell also works closely with taste for the things we put in our mouths. The moisture in our mouths increases the amounts of gas emanating from food. What we sense as flavor is really a combination of taste and smell. Without a sense of smell, we would have trouble detecting the difference between common foods. (For the very large number of Americans without a sense of smell, often lost to smoking, this sort of trouble is not at all hypothetical.)

Big Background: How Smell is Different

When sensation is taught in Introductory Psychology classes, there is usually a focus on vision. In part, this is justified in that there are important similarities among the different senses. The sense receptors fall into separate categories: monochromatic and the three additive colors in vision; the various frequencies in hearing; hot, cold, pressure, and pain in touch; and sweet, salt, sour, and bitter in taste. Signals from the sense organ go first to a specialized brain center for that sense and are then transmitted to the polysensory cortex. The various dimensions of the sense derived from the different types of sense receptors are retained and augmented throughout this transmission. And so on.

None of this is true of the sense of smell. No dimensions or basic categories from which all smells are derived have yet been discovered. The sensory nerves from the nose feed directly into the forebrain, the purported location of conscious, deliberate thought and so-called executive decision-making. There is no smell center in the brain akin to the vision or hearing centers. There is no evidence that the polysensory cortex gets involved. The rules learned from the study of vision simply do not apply.

Evolution can explain this by the fact that smell is a much older sense than the others. However, this explanation does not help us understand how smell works. In humans, vision is the most important of the senses in that it provides more information and uses more brain resources than the other senses. By the same logic, smell is the least important sense in humans. Many people become anosmic in adulthood, lacking a sense of smell almost entirely. (Cigarette smokers are almost all anosmic.) Unlike the blind or the deaf, the anosmic are able to get around in modern technological society not only unimpaired, but even undetected. We don't really know how many of us are anosmic and to what degree.

It is no accident then that psychologists have spent much more time studying vision than they have studying smell. All present-day psychologists are human. It is, however, unclear that the study of vision gives us the best picture of how sensation serves the psychological functions leading to behavior. It certainly gives us the best picture of how most sensation affects *human* behavior, but that may be more a matter of evolutionary accident than anything else. For other animals that psychologists use to study behavior, such as rats, it is the sense of smell that is most important. Had rats evolved into highly intelligent creatures (or, at least, intelligent enough to become psychology professors), research in sensation might have just as easily become much more focused on smell than on vision.

Summary

Sensation and perception are closely related, which is why some textbook authors choose to combine these two topics. Sensation is the process whereby stimulation of receptor cells sends nerve impulses to the brain, where they are registered.

Perception, which we cover in the next chapter, is the process whereby the brain interprets sensations, giving them order and meaning.

Quiz

1. _____ have developed mathematical models to describe the relationship between what is real and what is experienced.

    (a) Psychophysicists

    (b) Anatomists

    (c) Physiologists

    (d) Sensationists

2. The _____ is the smallest amount of energy that makes for a detectible sensation.

    (a) Just noticeable difference

    (b) Weber's Law

    (c) Absolute threshold

    (d) Signal detection theory

3. _____ states that a constant percentage of a change in the magnitude of a stimulus is necessary to detect a difference.

    (a) Signal detection theory

    (b) Weber's Law

    (c) Absolute threshold

    (d) Just noticeable difference

4. In signal detection theory, when a signal is reported as being perceived but, in reality, is not present, it is classified as a _____.

    (a) Hit

    (b) Miss

    (c) False alarm

    (d) Correct rejection

5. _____ are primarily responsible for color vision, and _____ are primarily responsible for night and dim-light vision.

    (a) Rods, cones

    (b) Cones, fovea

    (c) Rods, optic nerve

    (d) Cones, rods

6. Which theory of color vision cannot explain afterimages?

    (a) Trichromatic theory

    (b) Retinex theory

    (c) Opponent-process theory

    (d) None of the above

7. Which theory of color vision explains color vision at the bipolar and ganglion cell level?

    (a) Retinex theory

    (b) Opponent-process theory

    (c) Trichromatic theory

    (d) All of the above

8. In hearing, _____ theory is believed to be true for low-pitched sounds and _____ theory is true for high-pitched sounds.

    (a) Retinex, frequency

    (b) Vibration, pitch

    (c) Pitch, frequency

    (d) Frequency, pitch

9. The senses that relate to location are called the _____ senses.

    (a) Cutaneous

    (b) Vestibular

    (c) Somasthetic

    (d) Kinesthetic

10. The senses of taste and smell are categorized as the _____ senses.

    (a) Somasthetic

    (b) Vestibular

    (c) Kinesthetic

    (d) Chemical

CHAPTER 16

The Study of Perception

In the previous chapter, we discussed how signals enter the various sensory systems from the outside world and are registered by the brain. In this chapter, we will see how sensory information is organized into a form that can be used to understand the world and cope with it effectively.

Overview

Sensation converts the physical form of the incoming signals from light, air pressure, mechanical pressure, skin temperature, chemical vapors, and so on, into neural signals. But there is still more processing to be done. At the point where they impact the sense organs, a signal's logical form is not organized to tell us what is going on in the world. One physical object can make many differently shaped patterns of light on the retina depending on how our head is turned. Multiple sounds, smells, and tastes overlay one another in the ear, nose, and mouth because they happen at the same time. *Perception* is the name used by psychologists to describe how

the logical forms of sensations are transformed into something useful after they impact the nervous system.

HISTORY

Since people started wondering how people operate, it has been obvious that the sense organs deliver information about the world to somewhere inside the body. Plato talked about sensation, and his ideas are remarkably similar to those of Buddha, who preceded him. However, the notion that both the physical and the logical form of this information must be transformed, and may be transformed separately, is a relatively new idea. In the West, it was not until the eighteenth century when the philosopher, Thomas Reid, separated out what we now call perception from sensation.

CURRENT APPROACHES

An important focus of perceptual research has been on the basic processes that may serve as the building blocks for the necessary logical transformations that occur. Some of these processes appear in more than one sensory system. In addition, because the result of perception is meaningful information, important research focuses on how meaning is attached to sensory information. Finally, important facts about how perception works can be uncovered by determining when and how perceptual systems fail. This gives *perceptual illusions* an important role in research on perception.

It is important to understand the distinction between sensation and perception. Think of sensation as the gathering of raw data about the world. We receive signals through our visual system, our auditory system, and so on, and our brain acknowledges that we received the signals. Perception, on the other hand, is how we turn this raw data into useful information about the world. Our brain interprets the incoming signals, combines them with other signals from other sensory systems, and organizes them so that we can interact with the world. Without both processes, much (if not all) of the human experience would be lost to us.

Another important distinction between sensation and perception is that sensation is similar from person to person, but perception can vary widely. Given a normal sensory system, signals move through the system to the brain in essentially the same way. Thus, photons reflecting off this page travel through your visual system into the brain in roughly the same way as they do for every other human on the planet with a normal visual system. However, how those signals are organized and interpreted by the brain can vary widely from person to person. This is how you and a friend can see the exact same picture and perceive it differently.

Handy Hints: Speciesism in the Study of Perception

In this chapter, we focus on the study of vision because that is the primary focus of research in perception and where most of the examples can be found. It is no accident that human psychologists spend more time studying vision than the other senses, even when their concern is with perception, which is not fixed to any one sensory system. The vast majority of human brain resources devoted to sensation and perception are devoted to vision, as that is the sense we humans rely on most. Many of the laboratory animals used in these studies, however, rely mostly on other senses, such as the sense of smell. It could be argued that, when studying the rat, the best analogy to human vision would be to study the rat's sense of smell, but that is rarely what human psychologists do.

It is difficult to overstate the importance of the study of perception to the understanding of psychology. Everything we experience from the outside world is filtered through the lens of our perceptual processes. Although each sensory system has a corresponding perceptual system, we will focus on visual perception in this chapter. Visual perception has been the focus of much research, and therefore we have a more complete picture of the psychology behind visual perception. However, the principles involved in perceptual organization and integration are largely the same for all of the senses.

Perceptual Organization

Perceptual organization is the first step in the perceptual process. It is the process of integrating individual sensations into the framework of building blocks that make up our conception of the world around us: space, objects, motion, and so on. The flashes of light, wafts of scent, and bursts of noise, etc., that intrude on our senses do not make sense until they are fit into the structure of the environment we assume surrounds us. In this section, we will examine the perception of forms, depth or distance, and constancy. These are three ways that perceptual psychologists have characterized links between visual sensations and the structure of the world.

FORM PERCEPTION

Form perception involves organizing visual sensory information into meaningful patterns and shapes. Visual information arrives at the retina in the form of little colored dots, similar to the pixels on a computer screen or printer, laid out in two-dimensions. *Patterns* and *shapes* are words we use to describe the ways of organizing colored dots into larger two-dimensional elements.

Topical Talk: Figure-ground Separation

Figure-ground separation also plays an important role in auditory perception. Our brain is constantly registering auditory signals, yet we do not always pay attention to each sound we hear. Imagine that you are sitting in a crowded restaurant. Your brain is registering the sounds of the other diners, the background music, the silverware against the plates, your stomach grumbling, and other sounds you would expect to hear at a busy restaurant. Your brain is also registering the sounds of the air-conditioning (or heating) unit, the traffic outside, the cell phone across the room, and sounds you do not necessarily expect to hear. Yet, with all of these auditory signals being registered by the brain, you are still able to clearly hear your waiter describing the evening's specials. You do this by separating out the waiter's voice as the figure and the rest of the signals as the background.

Do you think that the figure-ground distinction applies to touch or taste or smell?

When something is in our visual field, the first step is to separate the *figure* from the *ground*. The figure is the object we are viewing, and the ground is the background behind the figure. For example, when looking at this page, you separate the black letters as the figure from the white background of the paper. This may seem obvious. After all, everyone familiar with books would separate figure from ground in this way. To demonstrate that this is, indeed, a psychological phenomenon (as opposed to a property of pages with printed letters), take a look at the classic "vase" reversible figure (Figure 16-1). The sensory information coming in to your visual system is the same as for any other person. However, how you separate figure from ground determines how you perceive the picture. If you separate the white as the figure and the black as the ground, you perceive a white vase on a black background. If you separate the black as the figure and the white as the ground, you perceive two faces in profile.

Figure 16-1   A white vase?

Tips on Terms: The Whole Is More Than the Sum of Its Parts

The "slogan" for the Gestalt school of psychology is, "The whole is more than the sum of its parts." The Gestalt psychologists believed that it was impossible to study the perception of forms by simply analyzing each of the sensations registered by the brain. They argue that it is often the case that perceptions are more than the sensations that give rise to them. That "more" is the meaningful pattern, or the whole. Thus, what we perceive (the whole) is made up of more than just the sum of the sensations (the parts).

Once the figure has been separated from the ground, several other perceptual organization processes can come in to play. These are commonly known as the *Gestalt principles of perceptual organization* and are the rules the brain uses to organize sensory information automatically into meaningful two-dimensional wholes.

The Gestalt psychologists proposed general rules for how the brain organizes sensory information to perceive forms:

- **Closure**   When a figure is interrupted or incomplete, we tend to close or complete the figure. For example, we perceive Figure 16-2(a) as a triangle, even though the sensory stimuli is three lines at various angles.

- **Continuation**   Stimuli that fall along a straight line or smooth curve are perceived as being grouped together. For example, we perceive the top portion of Figure 16-2(b) as a single line instead of four small lines. However, we perceive the bottom portion of part (b) as two groups of lines because they do not all fall along the same line or curve.

- **Proximity**   Stimuli that are close together are perceived as being grouped together. For example, we perceive the top portion of Figure 16-2(c) as three groups and the bottom portion as two groups, even though each row has six stimuli.

- **Similarity**   Stimuli that are similar to each other are perceived as being grouped together. For example, we perceive the top portion of Figure 16-2(d) as two groups, while the bottom portion of part (d) is perceived as four groups, even though the stimuli are the same (four stars and four triangles).

- **Simplicity**   We tend to perceive the simplest pattern possible. For example, ( ) is perceived as a set of parentheses instead of two single parentheses, whereas)(is not.

The Gestalt principles demonstrate that the brain organizes perceptual experiences in a way that reflects how things are typically organized in nature. When we

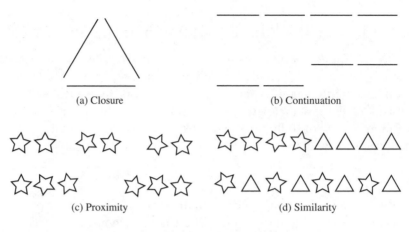

(a) Closure   (b) Continuation

(c) Proximity   (d) Similarity

Figure 16-2   Four Gestalt principles

observe a cat sitting in front of a tree, we do not think that there is a cat-shaped hole in the part of the tree that is obscured. We close the figure (principle of closure) and perceive the tree to be complete. Thus, in spite of the signal from our visual sensory system in which part of the tree is missing, we perceive the tree to be complete. From an evolutionary perspective, this shows that our brain evolved to pick up these cues and organize incoming sensory information accordingly. From a learning perspective, experience teaches us that trees are common, but that trees with cat-shaped holes are not.

More generally, experience teaches us that the portion of a large object hidden behind a small one usually looks like the unhidden parts. Whether due to evolution or learning, our perceptual system organizes information in ways that work to make us effective in the world. That our brain organizes nonnatural stimuli, such as the examples in the previous list and shown in Figure 16-2, according to the Gestalt principles demonstrates that our brain operates at this high level of abstraction and not with respect to prior experience (either evolutionary or learned) with specific types of objects.

DEPTH PERCEPTION

Because we live in a three-dimensional environment, it is important for us to be able to perceive the depth of objects in our visual field. *Depth perception* is the interpretation of visual cues that indicate how far away or how near objects are. What makes depth perception a particularly interesting topic of study for psychologists is the fact that the perception of depth or distance is not a function of the structure of the eye. The visual sensory system is incapable of registering signals of depth because the image projected on the retina is only in two dimensions (height and width). Thus, any perception of depth or distance is solely a function of how the brain organizes various cues in the environment. Depth perception is a function of the brain, not the eye.

In order to perceive depth, a person must interpret cues from the environment. There are a wide variety of environmental cues, but we can classify them into two groups: monocular cues and binocular cues.

Monocular Cues

Monocular depth cues are those cues that require only one eye to be perceived. Both eyes can pick up monocular cues, but only a single eye is required. Many monocular depth cues are considered *pictorial cues* because they are used in two-dimensional pictures to convey three-dimensional space. When you look at a painting or photograph, you perceive depth even though the picture is a flat surface. The following are some of the more common monocular depth cues:

- **Relative size**   When you think that two objects are roughly the same size, the object that casts the smaller retinal image is perceived to be more distant.

- **Closeness to horizon**   Objects that are closer to the horizon line are perceived to be more distant.

- **Linear perspective**   Parallel lines that converge toward the top are perceived to be converging into the distance.

- **Texture gradient**   In a highly textured scene, near objects appear coarser and more distant objects appear smoother. It is assumed that you cannot make out the details of objects that are farther away.

- **Interposition**   When one object appears to cover another object, we perceive the covered object to be the more distant.

- **Familiar size**   Because we assume that an object is its usual size, a familiar object that appears smaller is perceived to be more distant.

- **Motion parallax**   An object moving slowly across our visual field will be perceived to be more distant than an object moving more quickly across our visual field.

It is important to note that motion parallax is not traditionally considered to be a pictorial depth cue because photographs and paintings do not contain motion. However, another two-dimensional graphic medium lends itself quite well to motion parallax. Computer and video games have long relied on motion parallax to convey depth. A character in the "foreground" will appear to be moving across the screen faster than the objects in the "background." This cue tricks our mind into perceiving depth where there actually is none.

This is another demonstration of how our brain has evolved to pick up cues that are found in nature and even apply those cues in a nonnatural environment. It is true that when one object covers another, the obscured object is the one that is more distant. However, when we view the same two objects in a painting, we still perceive depth even though none exists. It is precisely because depth and distance work this way in nature that our brain evolved to register those cues. Our brains are so adept at picking up those cues that we even apply them to situations where no depth actually exists, such as in cartoons.

Binocular Cues

Binocular depth cues require both eyes to be perceived. These cues are based on the fact that our eyes are set physically apart from each other, and thus the image cast by an object on each retina is slightly different. To demonstrate this *retinal disparity*, hold your finger several inches from your nose and close one eye at a time. Notice how your finger seems to jump slightly to the left or right depending on which eye you use? This is the difference in the image cast on each retina. Now line your finger up with an object off in the distance and alternately close each eye again. Notice that the displacement of your finger is greater than the displacement of the object in the distance. We use this retinal disparity to determine depth in the visual field and to judge distance.

A second type of binocular depth cue is the *convergence* of the eyes to bring an object into focus. If you look at an object 20 feet in the distance, you'll notice that your pupils are essentially parallel to each other. However, if you look at the end of your nose, you'll notice that your pupils have to turn inward toward your nose in order to focus. We use the degree of convergence to determine depth in our visual field. The closer an object is, the more our eyes have to converge toward the object. Conversely, the farther an object is, the more parallel our pupils have to be. Our brains are wired up to detect the relative angles of our pupils to help inform us about the distance of objects.

PERCEPTUAL CONSTANCY

We have seen how our perceptual system deals with issues in two and three dimensions. But what about the fourth dimension? What about time? Ordinarily, when we think of objects in time, we think of motion. Changes in the location of the object over time will give us a sense of motion, and there are Gestalt principles to deal with that as well. However, it turns out that there is something more basic than motion involving time. Over time and at different times, the effect that an object has on our senses differs, even though the object itself is unchanged. In order to make sense of our world, we need to ignore all of the unimportant changes.

Objects are not perceived to change size, shape, or color, even when we view them from different distances, perspectives, and lighting situations. Our brain perceives the characteristics of objects to be stable regardless of the sensations registered by the visual system. This *perceptual constancy* enables us to function in a world in which familiar objects would appear to our sensory systems to be constantly changing. Instead of treating familiar objects as novel objects each time we encounter them in differing conditions, our brain compensates automatically for the sensory information.

There are three primary types of perceptual constancy in the visual system. *Color constancy* refers to the fact that we do not perceive that the color of an object changes when we view it under different lighting conditions. For example, you do not perceive that your white cat suddenly turns gray just because you turn off the light. *Shape constancy* refers to the fact that we do not perceive that the shape of an object changes when we view it from a different perspective. To continue the example, you do not perceive that your cat changes shape when you view it from above or from the side. Finally, *size constancy* refers to the fact that we do not perceive that the size of an object changes when we view it from different distances. You do not perceive your cat shrinking in size when it is walking away from you. In each of the above cases, the retinal image cast by the cat is different, yet the brain has compensated for those differences, and you perceive the size, shape, or color to be stable.

Perceptual Interpretation

Perceptual organization is just the first step in processing sensory information. The second step is to interpret the information. *Perceptual interpretation* is the process of giving meaning to sensory input. It is not enough to perceive that the object casting an increasingly larger retinal image is actually an object moving toward you; it is also important to know whether that object is your cat or a car.

Tips on Terms: Perceptual Meaning

When psychologists speak about *meaning* in the context of *perception*, they are talking about the identity of objects. We organize our perceptions into separate objects, and then we interpret our perceptions by figuring out what sorts of objects they are.

The world of meaning involves both natural and nonnatural phenomena. Each of us knows about different aspects of the world in terms of our own experience. Each of us understands the world differently in terms of our own needs and desires. Therefore, perceptual interpretation can vary greatly from person to person. In this section, we will examine how our experiences, our expectations, and our motivation influence our perceptions.

EXPERIENCE

Does our experience with the world influence how we perceive sensations, or do we build our perceptions based on the sensations registered by our brain? Two competing theories attempt to answer this question: *bottom-up* processing and *top-down* processing. Bottom-up processing theorizes that perception flows from sensory information up to the brain. Our brain forms perceptions by combining various sensory inputs until we understand what we are seeing (or hearing). The sensory information acts as the raw data from which perceptions are built. For example, if we see a nose and then an eye and then a beard, we combine this sensory information to perceive a man's face. Top-down processing, on the other hand, holds that perception begins with our experience and knowledge of the world. To form a perception, we use our prior knowledge to organize and interpret sensations as soon as they are registered by the brain. In this case, we may perceive the beard first and expect to perceive the rest of the man's face. Although these were once considered to be competing theories, contemporary research demonstrates that we actually perform *both* processes simultaneously.

EXPECTATION

Expectations play a large role in the top-down processing of sensory information. Our experience and prior knowledge of the world allows us to expect to perceive certain things in certain situations. These expectations create a *perceptual set* that can make some perceptions more likely to occur than others. Two important influences on our perceptual set are the external context and our internal schemas. The perceptual context is the external situation in which a person perceives something. For example, imagine that you overhear someone say, "That toast was cold." If the context is breakfast, then you perceive the statement to mean that the crisply heated bread served with the meal was cold. If the context is a wedding reception, you perceive the statement to mean that one of the well-wishers did not convey warm feelings when she raised her glass and voiced her sentiments. It is the external context that determines how you perceive the statement.

Our internal schemas also shape our perceptions. We discussed schemas in greater depth Chapter 14, but it warrants a brief mention here. Schemas are organized patterns of knowledge of objects, people, and situations. They allow us to make our environment predictable. In the previous example, our schema of a wedding reception contains information about who is supposed to be there, what is supposed to happen, and the order in which things are supposed to happen. Based on your schema, you know that the toastgiver is supposed to say warm things about the bride and groom. Because your schema provides this expectation, you may remember the toastgiver saying warm things, even when she did not.

MOTIVATION

Motivation is also a powerful influence on top-down processing in perception. Strong motivations can lead us to perceive something that is not present, to not perceive something that is present, or to perceive something that we might not otherwise notice. For example, imagine a man who is extremely thirsty. In the first case, the thirst may motivate him to see a mirage of an oasis in the desert. He perceives something that is not actually there. In the second case, he may drink an entire glass of a clear liquid without perceiving the smell of gin. He does not perceive the odor that is there. Finally, our thirsty man may notice glass vases of flowers in the hotel lobby. He perceives something he ordinarily wouldn't notice. In each case, the motivation of thirst influences what is perceived and how it is perceived.

Summary

Perception, like memory, is a part of the mind or brain psychologists have never seen directly, but are forced to assume exists due to the complexities of behavior. In the case of memory, we see relationships between present and past behavior, and memory is the hypothetical construct that fills the gap in time. Because we have external devices and tools, such as filing cabinets, marks on paper, and computer disks, that also bridge gaps in time, psychologists imagine that memory is somehow similar to these various devices. Perception, on the other hand, is a process, not a place, and we have very few tools that do anything similar. We do not have any convenient models or analogies to describe perception, but we study it anyway.

How do we know that perceptual processes exist? We understand the physics of how the senses operate, and we know that people operate effectively in a world organized very differently from the organization of the sensory information. People operate in a world of objects, space, time, and motion. They cope with the organization inherent in those sorts of things. But the light that hits the eyes is organized into little flashes across the retina. The sounds that hit the ears are organized into filaments in the cochlea vibrating at different rates. Touch is broken down into temperature and pressure and pain. Taste and smell involve mysterious chemical reactions on the tongue and in the nose.

Objects in motion over time and through space may cause all these sensory effects, but by the time the effects reach our skin, they have been completely transformed. And then, using our senses, we pick up some part of these effects, in terms of *their* effects on our senses. If the effects reach our skin, we can learn about the temperature, but not the color. If the effects reach our eyes, we can learn about the color, but not the temperature. And so on.

And, after all of the effects of the objects in the world have registered on our brains, how do we make sense of it all? How do we put all those puzzle pieces back together so that we know what is going on around us in terms of objects and space and time and motion, etc.? The name psychologists give to the process of putting everything back together is perception. And all we know about perception is in terms of the effects we have observed. After all, as we learned in our study of Structuralist psychology in Chapters 1 and 14, humans are not able to see how their own perceptual processes work, much less report on them reliably.

Finally, how do we know that perception is not just a part of the sensory processing? Logically, the fact that many of the things in our world can be seen *and* felt *and* heard *and* tasted *and* smelled strongly suggests that some perceptual processing is not tied to the separate senses. When we see a large object hit the ground with a loud bang, it all fits together automatically. We don't go searching for a source for the noise or a noise to fit the impact. It has all been integrated together for us.

Suppose someone said to us: "Helen Keller was an extremely perceptive person." Would this make sense to us despite the fact that Keller was blind and deaf? When we say she was perceptive, we are not saying that she smelled effectively or touched carefully. We mean that she had a very clear notion of what was going on in the world. Helen Keller was able to get that very clear notion despite not being able to use two of her most important senses. She perceived the world quite differently than most of us, but very effectively nonetheless.

Quiz

1. _____ can vary greatly from person to person, but _____ is essentially the same from person to person.

    (a) Sensation, perception

    (b) Perception, sensation

    (c) Illusions, perceptions

    (d) None of the above

2. What is the "slogan" for Gestalt psychology?

    (a) The hole is more than some of its parts.

    (b) The whole is more than some of its parts.

    (c) The hole is more than the sum of its parts.

    (d) The whole is more than the sum of its parts.

3. What is the first step in form perception?

    (a) Closure

    (b) Figure-ground separation

    (c) Proximity

    (d) Similarity

4. Which Gestalt principle states that stimuli that fall along a straight line or smooth curve are perceived as being grouped together?

    (a) Simplicity

    (b) Closure

    (c) Continuation

    (d) Proximity

5. The ability to perceive depth is a function of the _____.

    (a) Brain

    (b) Eye

    (c) Both a and b

    (d) None of the above

6. _____ depth cues are those that require one eye in order to be perceived; _____ depth cues are those that require both eyes in order to be perceived.

    (a) Binocular; monocular

    (b) Monocular; binocular

    (c) Size; linear

    (d) Linear; size

7. Which of the following is not a type of perceptual constancy?

    (a) Color constancy

    (b) Shape constancy

    (c) Size constancy

    (d) Motion constancy

8. _____ processing theorizes that perception flows from sensory information to the brain.

(a) Top-down

(b) Lateral

(c) Bottom-up

(d) Depth

9. In top-down processing of sensory information, our expectations create a(n) _____.

(a) Perceptual set

(b) External context

(c) Internal schema

(d) Perceptual constancy

10. Motivation influences top-down processing by leading us to

(a) Perceive something that is not there

(b) Perceive something that we might not otherwise have noticed

(c) Not perceive something that is there

(d) All of the above

CHAPTER 17

Neuroscience and Genetics

Psychology is about the behavior of the entire organism. However, it is clear that the answers as to *how* we are able to behave as we do are to be found in the nervous system, particularly in the brain. Biologists study brain anatomy and function. Psychologists try to understand how brain function makes behavior happen.

Critical Caution

Understanding how the brain works is very different from understanding how the brain makes behavior happen. Unfortunately, at the present time, we know astonishingly little about both. Although we know thousands upon thousands of things about the brain and nervous system, and more and more each day, we are really just scratching the surface. There is much, much more to learn.

Overview

Neuroscientists focus on the biological and physiological basis of behavior. On the experimental level, they research animal behavior, biochemistry, electrical activity of the brain, and brain anatomy. Eventually, links will be forged between what we know about the nervous system and what we know about behavior. In this chapter, we will focus on the basics of what is known about the nervous system. Drawing the connections between behavior and the brain is the subject of more advanced study.

HISTORY

To us, it is a given that the nervous system, and particularly the brain, are the most important parts of the body in terms of producing behavior. But this was not always the case. The ancient Greeks believed that the brain was a sort of radiator for cooling the blood. The liver was thought to be the main source of behavior, with the heart being the seat of the emotions and other organs serving other functions.

By the time of René Descartes (1596–1650), the importance of the nervous system and the brain had been recognized. There was even a simplistic notion of how the nervous system worked in order to produce behavior. This notion, for which Descartes was an early key advocate, was called the *reflex*. Simple reflexes, such as the patellar reflex, are a way that the *peripheral nervous system*—the part of the nervous system farthest from the brain (in the arms and legs, etc.)—works. With a simple reflex, a prod (in Latin: *stimulus*) to the sensory nerve endings tends to produce a *response* in the form of a muscle movement. For example, with the patellar reflex, the doctor hits your knee with a rubber hammer (the stimulus) and your foot jerks forward (the response).

Reflexology was the idea that the nervous system produces behavior using more complex versions of the reflex. Complex environmental events, for instance, books and elections and cars and other people, act as prods to produce complex behavior, for instance, reading and voting and driving to the beach and going out on dates. As we have learned more and more about the rest of the nervous system, it is clear that it is not structured in the same way as are the nerves in the arms and legs. It is not clear how thinking of all behavior as reflexes helps us to understand how the nervous system as a whole produces the many different kinds of complex behavior.

CURRENT APPROACHES

Traces of reflexology still persist in many areas of psychology. The old terms, *stimulus* and *response*, are still commonly used. In contrast, over the last 25 years, there has been an explosion of new discoveries about how the nervous system works. Three general approaches to understanding how psychological phenomena, often described using the older terminology, have arisen from the newly discovered activities of the nervous system.

The first approach is a *bottom-up approach*. As more and more details of how the nervous system works arise, researchers attempt to figure out exactly what the nervous system is doing at the cellular level. Then they attempt to tie these microscopic events to details of psychological phenomena. The biggest problem with this approach is that it is like trying to solve a jigsaw puzzle without knowing what the picture looks like. And we are missing a lot of pieces.

The second approach is *modular*. Specific parts of the nervous system act almost independently as subsystems. The visual system is a good example. For instance, our eyes respond to moving stimuli by following the motion. The overall pattern of the behavior is like a reflex, but circuits deep in the brain are involved. If those brain circuits can be traced, then the entire subsystem can be understood. This is like trying to solve one part of the jigsaw puzzle where there is an easily identifiable object pictured, such as a face or a bird or a window, and where we have most of the pieces. This may or may not be helpful in solving the entire puzzle later on, but we will have solved that portion.

The third approach is called *cognitive neuroscience*. In this approach, cognitive theory is used to describe the psychological function behind some behavioral phenomenon. Then the neuroscientific data is examined to find neural mechanisms that might make the cognitive theory work. This is like knowing that the jigsaw puzzle is of a famous painting and guessing which painting it is. The only problem is that, if the cognitive theory is wrong, the entire approach may lead nowhere.

The Nervous System

The nervous system is the body system most critically involved in producing behavior. Like the other body systems, it is distinguished by the types of tissue it contains and the functions it performs. The nervous system is composed of nervous tissue, which is composed of two main types of cells: the *glia cells* and the *neurons*. The glia cells are found in the brain. Neurons are found throughout the nervous system.

The nervous system is divided into two parts: the *peripheral nervous system* and the *central nervous system (CNS)*. The CNS is defined as the brain and the spinal cord. The peripheral nervous system is all the rest.

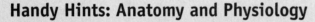

Handy Hints: Anatomy and Physiology

Biologists divide up bodies in two ways, anatomically and physiologically. Anatomical divisions are based on structure. Different parts are found in different places, usually physically separated from one another or made up of different types of tissue (or both). They usually have different shapes. Physiological divisions are based on function. The different parts have different roles and serve different purposes for the animal as a whole. The nervous system is distinguished from the rest of the body both by structure and by function. The distinction between the peripheral and central nervous systems is based on anatomy. While some functions are found only in one part or another, the nervous system works as a whole, across the division between the periphery and the center.

PERIPHERAL NERVOUS SYSTEM

The peripheral nervous system is composed of long bundles of neurons, called *nerves,* that connect different parts of the body. The peripheral nervous system is divided up into parts:

- **Skeletal (somatic) nervous system**   The nerves that connect the nervous system to the parts of the body that interact with the world.

  - **Sensory nervous system**   The nerves that relay information from the senses to the nervous system, particularly to the brain.

  - **Motoric nervous system**   The nerves that relay information to and from the skeletal muscles, which move the body.

- **Autonomic nervous system (ANS)**   The nerves that communicate with and control the body systems that manage basic life functions such as respiration, circulation, reproduction, and digestion.

  - **Sympathetic nervous system**   The portion of the autonomic nervous system that speeds up respiration and circulation and slows down digestion, enabling the animal to deal with stressful and/or dangerous situations.

  - **Parasympathetic nervous system**   The portion of the autonomic nervous system that slows down respiration and circulation and speeds up digestion, returning the body to its more normal state.

The nerves of the autonomic nervous system connect directly to glands and also connect to the smooth muscles. Smooth muscles are the muscles that operate the heart and lungs and digestion. They are distinguished from the skeletal muscles, which are controlled by the skeletal nervous system, are connected to the bones, and move the body.

Tips on Terms: Afferent and Efferent

Afferent is a term referring to peripheral nerves that carry signals inward toward the central nervous system (even if they don't get all the way there). *Efferent* is a term referring to peripheral nerves that carry signals outward, away from the central nervous system.

Skeletal Nervous System

The sensory nervous system was discussed in Chapter 15. All the nerves of the motoric nervous system transmit signals in two directions. Not only do they carry the signals that cause the muscles to move, but also they carry the sensory signals about how the muscles moved and where the body is now. The role of the skeletal (somatic) nervous system, specifically the motoric nervous system, in moving the body is twofold: Some movements are controlled by the skeletal nervous system without involving the brain. These are the reflexes, mentioned previously. The other role of the skeletal nerves is to transmit signals to and from the brain, making deliberate action possible through muscle movements.

In the case of the simple reflex, the stimulus causes afferent nerve impulses from the muscle to travel via a long nerve to the spinal cord. In the spinal cord, the impulses are transferred to an efferent nerve. The impulses then travel back to the muscle, causing it to contract. In more complex reflexes, nerves from higher up in the nervous system, including in the brain, send signals to the juncture in the spinal cord, affecting its action.

Autonomic Nervous System

The autonomic nervous system enables the "four *F*s": fighting, fleeing, feeding, and sexual behavior. The sympathetic nervous system prepares us for fighting, fleeing,

Variant View: The Sensory Nervous System

Most introductory psychology texts do not list the sensory nervous system as part of the peripheral nervous system. In part, this is because psychologists treat sensation as a separate topic. It is also because the nerves that carry afferent signals about muscle movement are the same nerves that carry the efferent signals that cause muscle movement. These afferent signals are sensory signals, but the nerves are part of the skeletal (somatic) nervous system. On the other hand, all of the senses are connected to the nervous system by afferent nerves. Clearly these nerves are part of the peripheral nervous system.

Variant View: Glia Cells

For decades, it has been assumed that glia cells do not contribute directly to the neural activity that underlies behavior, but only indirectly by supporting neuronal activity. However, glia cells do pass signals through the nervous system, and there is no guarantee that these signals are not involved in behavior.

and sexual behavior. The parasympathetic nervous system enables feeding and brings sexual behavior to completion. The effects of the sympathetic nervous system tend to operate on all of the organs and body systems simultaneously. The effects of the parasympathetic nervous system tend to operate on one or only a few organs or systems at a time.

CENTRAL NERVOUS SYSTEM

The central nervous system (CNS) is made up of the brain and spinal cord. Neuroscientists are interested in looking at behavior in terms of the electrochemical processes that occur in the CNS. The glia cells are the more numerous and smaller of the two cell types. They surround the neurons and hold them in place. Glia cells also carry nutrients to the neurons, remove waste products, and act as a barrier against harmful substances that may get into the blood. Thus, glia cells are important in helping the neurons function.

The brain is considered the most complex organ in the body, and is the centerpiece of the nervous system. The brain works as a unified whole, but neuropsychologists can identify which structures perform which functions. The brain is organized into three interconnected layers, each of which contains structures that regulate everyday life.

Big Background: Which Part of the Brain Does What?

Figuring out which part of the brain does what is no easy matter. Before modern brain-scanning techniques, the most common method was to observe people with damage to particular parts of the brain or to damage specific parts of the brains of nonhuman animals. This is why the descriptions of brain parts that follow will include information on brain damage. The biggest problem with this method is that it is hard to find people with damage to exactly one part of the brain.

Central Core

The central core of the brain is found in all vertebrates. Its main structures are important for regulating the basic processes that keep us alive. These structures also perform the early stages of processing sensory information.

The *thalamus* begins processing sensory information. It registers the information, determines its fundamental properties (such as whether it is good or bad), and sends it to the appropriate area of the cortex. For example, the thalamus receives a signal for the ear indicating that a sound is incoming. It registers that it has received the signal. Next, it determines whether the sound is good or bad, such as whether it is a police siren or an ice cream truck. Finally, it sends it to the temporal lobe for further processing. Because the thalamus is important for sensory processing, damage to this area is often lethal.

The *pons* is involved in controlling sleep and arousal. Because of its importance, damage to this area is also lethal.

The *cerebellum* is involved in maintaining a sense of balance and equilibrium. It is also involved with coordinating muscle movements to ensure that they are smooth and efficient, and executing simple motor tasks. Damaging this area can result in a condition known as *ataxia*. Ataxia is characterized by a lack of balance, uncoordinated movement, and severe tremors. Ataxic patients lack the control for even simple movements, so they may accidentally punch you in the stomach while reaching out to shake your hand.

The *medulla* is found at the base of the skull. It plays an important role in many automatic activities, such as circulation, breathing, chewing, salivation, and facial movements. Damage to this area can interrupt any of these processes, and severe damage can cause death.

The *reticular formation* signals the cortex to attend to an incoming signal. It arouses the forebrain when information related to survival must be processed. It permits sustained attention and helps screen out extraneous sensory input, particularly during sleep. It can also exaggerate or amplify internal or external stimuli. Damage to this area can result in permanent coma.

Limbic System

The limbic system exists in all mammals. It mediates motivated behaviors, emotional states, and memory processes. It also regulates body temperature, blood pressure, and blood-sugar levels.

The *hippocampus* is a structure that is heavily involved in memory, movement, and spatial relations. If it is damaged on both sides of the brain, a person is not able to store new information. A person with this type of damage is not able to find her way to a familiar location, even though she has been there many times before.

Fun Facts: Sensory Deprivation

The reticular formation may play an important role in producing the bizarre experiences described during sensory deprivation experiments. When the brain receives very little input, as happens in a sensory deprivation situation, the reticular formation amplifies whatever information might be present. One person, who practiced a lot of sensory deprivation, donned a diving mask and immersed himself naked in a tank of lukewarm water. During this session, an air hose sprung a leak and small air bubbles occasionally floated up his leg. Under normal conditions, he probably would not have even noticed the bubbles. But, in a sensory deprivation situation, the reticular formation exaggerated their importance. Once out of the tank, he reported that the contact at first produced an experience of intense sexual pleasure. Later, however, when the leak worsened and the bubbles became more frequent, he reported that the intense pleasure turned into intense pain.

She is also not able to remember what she was doing or whom she was with moments before turning her attention to something else.

The *amygdala* is involved in learning and memory, particularly memories with emotional components. It helps link feelings with other experiences and is involved with emotional expression. It may strengthen perceptions that have emotional components, which may make them more memorable. Damage to this area in a lynx, a cat that is so wild it has to be handled with nets, makes the animal docile and tame. Damage to this area in humans can result in deficits such as failure to perceive threat and experience fear appropriately.

Case Study: Charles Whitman

Charles Whitman had been an altar boy and an Eagle Scout. When he became a student at the University of Texas, his life took a new direction. He assaulted his wife and became involved in several fights. Even more frightening, he confided to psychiatrists that he was fighting to resist urges to commit even more extreme violent behavior. One day, he lost that fight. He climbed to the top of the campus observation tower with a high-powered rifle and began shooting people. Whitman murdered 14 people and wounded more than 20 others before the police finally killed him. Further investigation showed that Whitman had also killed his wife and his mother. An autopsy on Whitman's body revealed a large tumor, about the size of a walnut, pressing against his amygdala. Although researchers cannot be sure that this tumor was the cause of his violence and aggressiveness, this area of the brain is known to regulate aggressive behavior in other animals.

Tips on Terms: Gray and White Matter

Gray matter refers to the portions of the brain consisting of a large proportion of cell bodies and dendrites. To the naked eye, it appears gray. *White matter* refers to the portions of the brain consisting of a large proportion of axons. It is lighter in color. Wherever white matter is found, it is presumed to be connecting two other regions of the brain, rather than being involved in brain processes at that location.

Cerebral Cortex

The cerebral cortex is the gray matter at the outermost area of the cerebral hemispheres. It directs the brain's higher cognitive and emotional functions. It is divided into two halves, or *hemispheres,* with each hemisphere having four matching lobes. The eight lobes come in pairs, with one lobe on the left and a matching lobe on the right. The areas in these lobes regulate all forms of conscious experience, including perception, emotion, thought, and planning. The cerebral cortex also regulates unconscious cognitive and emotional processes.

The *occipital lobe* is located at the rear of each hemisphere of the cerebral cortex. It plays a major role in vision in that it conducts the initial stages of analyzing visual information. Damage to this area can cause blind spots in the visual field.

The *parietal lobe* is located just forward of the occipital lobe. It is the receiving area for the sense of touch and is important for body perception. It integrates visual, auditory, and somatosensory information and passes it along to other parts of the cortex. It is also important for attention and language comprehension. People with damage to this area can feel a sensation on the skin and can see and hear, but they find it difficult to integrate these different sensations into a single perception of an object or event.

The *temporal lobe* is located by the temples. It is the area where auditory signals are first received and processed by the brain. It is also involved in complex aspects of vision, such as face recognition and storing visual memories. Because the temporal lobe connects with the limbic system, it may also be involved in emotional behaviors. Damage to this area can affect the ability to process auditory information, store visual memories, or comprehend language.

The *frontal lobe* is located in the front area of the cortex. It assists in motor control and cognitive activities such as planning, decision-making, and goal-setting. It seems to be involved in the ability to sort and order information, analyze steps leading to a solution, and anticipate future events. The frontal lobe is considered the seat of all higher mental functions, including language production. Damage to this area can negatively affect any of these processes.

Corpus Callosum

The *corpus callosum* is the large network of nerve cells that connect the two hemispheres of the brain. Its purpose is to facilitate communication between the two halves of the brain. When the corpus callosum is severed, it results in the *split-brain phenomenon*. This procedure was a surgical technique used to control severe epileptic seizures long ago. Psychologists noticed that there were several interesting behavioral results of severing the corpus callosum. They were able to study these patients to conduct important research on the specialized abilities of the brain hemispheres.

Sperry won a Nobel Prize for this line of research. He found that when a picture of an object is shown on a screen in half of the visual field, this information is transmitted only to the opposite hemisphere. In split-brain patients, the hemispheres are not connected and thus are independent of each other. If a picture of a spoon is presented in the right visual field, it is transferred to the left hemisphere. When the patient is asked to name the object, he can because the language centers are housed in the left hemisphere. If the same object is shown in the left visual field, and thus transmitted to the right hemisphere, the patient cannot name the object. This is due to the fact that the speech mechanisms are located in the left hemisphere. Even so, the right hemisphere "knows" the object and can instruct the left hand to pick out the spoon from a group of objects. This shows that visual-spatial tasks are handled by the right hemisphere of the brain.

NEUROENDROCRINE SYSTEM

The endocrine system is a group of organs, called the *endocrine glands*, spread throughout the body that release various chemicals that have a wide variety of effects on other body systems. The importance of the endocrine system from a behavioral perspective is that many of these chemicals are involved in the same sorts of regulatory activities that the autonomic nervous system is. In fact, many of the endocrine chemicals, called *hormones,* are also neurotransmitters (which you will read about in the next section on the neuron). Often, when these chemicals are released from endocrine glands, their effects on nonneural tissues are quite different from their effects on nerves. However, even when they are released from endocrine glands, as opposed to being released from neurons, they may come into contact with neural tissue and alter nervous system activity.

These are the most important of the endocrine glands, listed by their location in the body:

- **Head**   Two endocrine glands are located deep within the base of the brain.
  - **Pituitary gland (anterior)**   The master gland for the neuroendocrine system. It releases hormones that control the other glands. It is itself controlled by the hypothalamus in the brain.

- **Pituitary gland (posterior)**   Controls blood pressure and urine volume. Another hormone controls milk production in women.
- **Pineal gland**   Releases hormones that control sleepiness and the onset of puberty.
- **Neck**   Two glands are located at the front of the neck.
  - **Thyroid gland**   Controls metabolism.
  - **Parathyroids**   Regulates calcium and potassium levels.
- **Torso**   Two types of glands are located in or around the organs in the torso.
  - **Adrenal glands**   Located above the kidneys, these glands release epinephrine and nor epinephrine, which activate the sympathetic nervous system. Another hormone controls salt balance.
  - **Isles of Langerhans**   Located inside the pancreas, these cells control sugar utilization by cells.
- **Pelvis**   The sex organs that produce eggs and sperm are also glands.
  - **Ovaries (females)**   Regulates menstruation, pregnancy, and sexual behavior in women.
  - **Testes (males)**   Regulates sexual behavior in men.

THE NEURON

Neurons are the nerve cells. They are specialized to conduct signals from one part of the body to another. They are the fundamental building blocks of the nervous system. The basic function of any neuron is to receive information, process it, and transmit those impulses to other neurons, muscle cells, or gland cells.

Types of Neurons

There are three types of neurons. *Sensory neurons* convey information from the sense organs (eyes, ears, and so on) to the brain and spinal cord. *Motor neurons* convey information from the brain and spinal cord to muscle cells. *Interneurons* are neurons that connect other neurons and integrate the activities of the sensory and motor neurons. It is the relationship among interneurons that somehow becomes translated into thoughts, feelings, perceptions, memories, and ultimately, actions. In other words, they are involved in all the phenomena that interest psychologists.

Anatomy of a Neuron

Neurons vary greatly in size and shape, depending on their location in the nervous system. However, despite this diversity, neurons have three basic parts. The cell body,

or *soma*, is the nucleus of the cell. It provides the energy for all cellular activities. The *dendrites* are one type of fiber structure that branches out from the soma. They are usually short structures that have specialized areas for receiving impulses from other cells. The *axon* is the other fiber structure that branches out from the soma. It is relatively long and serves the function of carrying the impulse to the muscle cells, gland cells, or other neurons. All neurons have a soma and dendrites. Some CNS neurons lack axons.

Neural Communication

Neural communication is simply how the neurons relay messages to each other. This communication is based on the *action potential*. When a certain threshold of stimulation reaches the axon, the axon's membrane at the point of stimulation suddenly becomes permeable to sodium. The sodium channels are now open and the sodium ions are pulled into the cell by the negative charge on the inner surface of the membrane. For an instant, the inside of the membrane becomes positively charged relative to the outside of the membrane. This abrupt reversal in electrical charge across the cell membrane is the action potential. The action potential travels down the axon as sodium channels consecutively open down the length of the axon. The axon can quickly restore itself so it is able to conduct as many as 1,000 consecutive action potentials per second (although no neuron will sustain this rate of firing for very long). The speed at which an action potential travels down the axon ranges from 2 to 270 miles per hour.

Because neurons are not connected to each other physically, how does the action potential of one neuron stimulate another neuron? The gap between the end of the axon and the dendrites of another neuron or the cell membrane of another cell is the *synaptic cleft*. Taken together, the surfaces of the sending and receiving neurons on either side of the cleft, along with the cleft itself, are called the *synapse*. If the gap is small, the electrical current can flow from the presynaptic cell across to the cell

Fun Fact: The Myth of the Synaptic Spark Plug

For many years, it was believed that most synapses operated like automotive spark plugs without neurotransmitters. This was because most research on neural transmission had been conducted with invertebrates such as slugs and worms, which do not use neurotransmitters very much. In addition, the very, very few synapses in humans that do work like spark plugs are found mostly in the peripheral nervous system, where they are easier to get at and to study. The vast majority of human synapses rely upon one or more neurotransmitters. This is a good example of how research with nonhuman animals can lead us astray if we are not careful.

on the other side (post-synaptic cell). Most often, the gap is too large for the current to flow. In these cases, neural communication relies on neurotransmitters.

Neurotransmitters

Neurotransmitters are chemical substances stored at the end of axons that transmit signals to the postsynaptic cell by traveling across the synaptic cleft. When the action potential reaches the end of the axon, it stimulates the release of the neurotransmitter from the presynaptic cell. The neurotransmitter is released into the synaptic cleft and travels to the postsynaptic neuron. On the receiving neuron, there are receptor sites that are activated because the molecular structure of the neurotransmitter fits its receptors, as a key fits a lock. Thus, a specific neurotransmitter will activate only certain receptors and not others. This activation will have one of two effects, depending on the neurotransmitter and the receptor: *Excitation* occurs when the neurotransmitter activates an action potential in the receptor cell; *inhibition* occurs when the neurotransmitter causes the cell to pause.

What do these neurotransmitters do? It is relatively well established that different neurotransmitters affect different aspects of behavior. If you have ever heard that something was caused by "a chemical imbalance in the brain," you have heard of neurotransmitters. Researchers have discovered more than 75 different chemicals that can act as neurotransmitters. Table 17-1 lists the more common neurotransmitters and their behavioral effects.

Table 17-1   Neurotransmitters and Their Effects

Neurotransmitter	Involved In	Too Much	Too Little
Acetylcholine	Involved in attention, learning, memory, arousal, and motivated behaviors such as aggression and sexuality	Unknown	Memory impairment, possibly Alzheimer's disease
Dopamine	Thoughts, motivation, emotion	Involuntary movements, possibly schizophrenia	Parkinson's disease, memory, movement impairment
Norepinephrine	Mood, arousal	Anxiety, schizophrenia-like symptoms	Memory impairment, depression
Serotonin	Sleeping, arousal, mood, eating, pain	Unknown	Aggressiveness, sleeplessness, severe depression
GABA	Anxiety regulation	Unknown	Anxiety, neural excitement
Endorphins	Mood, pain regulation	Inhibits pain, runner's high	Increased pain sensitivity

Neural Complexity

The previous section describes the standard way that neurons communicate. However, there are many variations. The nervous system is extremely complex. Listed here are a few common misconceptions about neural communication:

- *One neuron can cause another neuron to fire.* Very unlikely. It is common for as many as 10,000 synapses to connect to the dendrites of a single neuron. For a single neuron to cause another to fire, it would have to be the only neuron connected to thousands of those synapses.

- *Neurons are generally at rest until stimulated.* True only for sensory and motor neurons. In the CNS, the large number of complex interconnections keeps most neurons firing at a moderate rate. Importantly, this means that information can be passed along by either increasing or decreasing the rate.

- *All neurons communicate via synapses.* False. In some cases, neurons release neurotransmitters across a wide region, potentially altering the activity of any synapses or neurons in the region. Because some endocrine glands release chemicals identical to neurotransmitters, this means that endocrine glands can alter neural activity.

- *Synapses always connect an axon to a dendrite.* False. Synapses can bridge the gap from an axon to a dendrite, or to a soma, or even to another axon.

- *All neurons transmit action potentials.* False. Action potentials are transmitted only along axons. All other neural signals are transmitted passively, like a spark down a wire. Neurons without axons cannot transmit action potentials.

- *All neurotransmitters are complex biochemicals.* False. The number of proven and suspected neurotransmitters keeps growing. Among them are some unlikely chemicals, including nitric oxide (NO) and possibly carbon monoxide (CO), which are toxic gases made up of tiny molecules and more commonly associated with automotive exhaust than neural activity.

- *Individual neurons release only one neurochemical.* Not always. Many neurons release more than one neurotransmitter, possibly into the same synapse. Not only can the same neuron participate in different types of neural activity, so can individual synapses.

- *There is only one type of receptor for each neurotransmitter.* False. A number of neurotransmitters affect different kinds of receptors. Dopamine affects at least four. The different receptors can be affected by other neurochemicals present in the synapse, altering the activity of the synapse.

Genetics

Genes determine the structure of the body, including the brain. In this sense, they have an enormous impact on behavior. Traditionally, however, the focus of genetic research in psychology has been on how individual genetic differences make different people behave differently from each other. Before the completion of the Human Genome Project, little was known about which genes varied from person to person. Therefore, studies on individual differences had to rely on statistical methods to attempt to relate overall genetic similarities with overall behavioral similarities for groups of people. Now there are a host of new methods for identifying which genes impact behavior and how.

Genes can have a number of effects at birth, such as producing different levels of various neurotransmitters and hormones and different numbers of the various types of neurotransmitters. In addition, genes have at least two other ways of influencing behavior. First, throughout gestation and into the first year of life, the brain grows. The brain continues to develop well into adolescence. Brain cells migrate from one part of the brain to another, leaving trails that become the axons that transmit signals from one part of the brain to another. Brain cells, and their connections to one another, grow and many also die. The final structure of the brain is determined by all of these factors. During this period, various genes communicate chemically with these processes and influence them. Second, even in the adult brain, certain chemicals, called *third messengers*, operate within the bodies of the nerve cells, traveling back and forth between the genetic nucleus and the synapses. The impact of third messengers on the activity of the synapses, and thus on behavior, is not yet well understood.

With the recent explosion of new knowledge in genetics, the impact of genetics on psychology is likely to be extremely profound.

Methods: Old and New

In at least two areas, genetic studies and viewing brain activity, recent advances are radically changing the way neuroscience and biological psychology are done. What will be discovered is anybody's guess. It is important, however, to take a look at the differences between the old and new methods in order to get an idea of what's coming.

MONITORING THE BRAIN

Traditional methods of looking to see what is going on in the brain have been very limited. A wide variety of new methods has radically expanded the ways we can look at brain activity and study it together with behavior.

EEGs and Individual ERPs

Until recently, the most common method for looking at brain activity was the electro-encephalograph or EEG. A series of electrodes is placed on the scalp at specific points. Electrical activity due to muscular activity, such as eyeblinks, is then cancelled out electrically. The remaining electrical activity is due to the brain. This technique works because, as discussed previously, action potentials are electrical signals and when large numbers of neurons are active, the resultant electrical activity can be measured at the scalp.

Using a variety of methods to separate out the various signals, different brain waves, such as the alpha and theta waves, can be measured separately. Also, coordinated activity in response to a particular stimulus can be averaged over repeated presentations of the stimulus, resulting in a recording known as an *event-related potential (ERP)*. The advantages of the EEG are that it is nonintrusive and relatively inexpensive. Disadvantages include the fact that it measures overall brain activity, without the ability to distinguish activity in different parts of the brain, and the fact that it is too slow to capture the most rapid neural activity.

Another method for recording ERPs involves inserting a single electrode into the brain through a hole in the skull until it begins to register a specific signal, which is assumed to come from a single neuron or small group of neurons. The advantage to the individual ERP is that it records highly specific information from a specific location in the brain. The disadvantages include the fact that it requires surgery (and is thus not appropriate for experiments on humans) and that it cannot capture information from more than one tiny area of the brain at a time.

Brain Scans and Multiple ERPs

In the past twenty or so years, a number of new techniques have been developed for examining the brain as it works. Sets of multiple ERP electrodes can now be inserted into the brain to record multiple sites in a limited area of the brain. Various scanning systems, such as Positron Emission Tomography (PET) scans, CT scans (a high-resolution X-ray), and functional Magnetic Resonance Imaging (fMRI), allow the entire brain to be examined in detail. Some of these methods allow the brain to be monitored as various behavioral experiments are being conducted. This is critically important because it allows researchers to relate behavior to the brain activity that goes on at the same time. Each of these methods has variants, and new variants are being invented as you read this. Each has its advantages and disadvantages.

The basic method currently in use is to perform some experiment, often one with a long history in psychology, while simultaneously scanning the brain. The pictures of the working brain during the performance of different behaviors by a single person allow researchers to see what parts of the brain are active only for one type of behavior. This allows researchers to obtain evidence about what parts of the brain contribute to what sorts of behaviors.

Modern brain-scan techniques monitor the entire brain for many seconds at a reasonably high speed at incredible levels of detail. Even so, it is important to note that researchers would like to be able to see things at an even greater level of detail, down to the level of individual neurons. They would also like to be able to monitor changes at a higher speed and for longer periods of time, not to mention while the person is walking around, rather than lying down inside a big machine. Improvements are desired, and they are coming.

GENES AND BEHAVIOR

The Human Genome Project, together with modern methods of genetic analysis, is rapidly rendering traditional methods of relating genes and behavior obsolete.

Nature Versus Nurture

The traditional study of genetics in psychology focuses on the influences of a person's genetic makeup on his behavior. This brings up the old philosophical question of nature versus nurture. It used to be that we questioned whether behavior was due to genetics or to the environment. However, it has been fairly well established that behavior is a function of both genetics and environment. The question has now become one of how a certain behavior is influenced by genetics and environmental factors.

Key Point: What Brain Scanning Can't Tell Us

Improvements in brain-scanning technology can only do so much. At present, brain scans tell researchers only where brain activity is going on, not what sort of activity is going on. One important caveat is that, as we have seen, neural signals can be either excitatory or inhibitory. When a researcher uses brain scanning to detect activity in a certain region of the brain, we have no way of knowing if that activity is excitatory or inhibitory. That region of the brain may be turning on or turning off. Brain scans are merely tools and need to be used as part of integrated research efforts in order to be effective.

Because most behaviors depend on the effects of several genes and their interaction with the environment, researchers have a particularly difficult job teasing out the genetic effects on behavior. However, there are several sources of research information.

Family studies examine blood relatives to determine how much they resemble each other on a specific trait. The problem is that blood relatives also share similar environments. Thus, heredity and environment are confounded. These types of studies often suggest hereditary contributions to behavior, but alone are not conclusive evidence.

Twin studies examine genetic influences by studying identical and fraternal twins. Because identical twins share the same genetic makeup, researchers begin with the assumption that any similarities in behavior are due to genetic similarity. Because fraternal twins do not share the same genetic makeup, they are less similar genetically than identical twins. Research has shown that identical twins are more similar than fraternal twins on IQ tests, anxiety, aggression, shyness, and many other behaviors. However, because identical twins are most often raised together and treated as a "set" of twins rather than individuals, the effects of genetics on behavior are confounded with environmental influences. It is impossible from this comparison to determine whether the similarities are due to similar genes or to similar upbringings and experiences.

To get around this confound, psychologists will often study identical twins separated at birth because they have not been raised in the same environment. Even though this is a good way to avoid confounding genetics and environment, it is rare to find a lot of these people. Although there have been astonishing similarities, psychologists cannot claim this as conclusive evidence because we do not know how much of this is a result of coincidence. However, fraternal twins that have been raised apart rarely show as many similarities.

Another important comparison can be found between adopted children and their biological parents and their adoptive parents. Studies have shown that adopted children are more likely to resemble their biological parents than their adoptive parents on factors such as alcoholism, intelligence, and personality factors. This indicates that these factors may have a strong genetic component, but the evidence from adoption studies is far from conclusive.

Alternative Alleles

The previous methods rely upon very broad notions of the percentage of genes shared between identical and fraternal twins, siblings, and unrelated people. They do not attempt to isolate or identify any individual genes or genetically controlled mechanisms. At best, they can only provide a statistic reflecting how much of a given measure of behavior is due to genetics and how much is due to environmental influence. This statistic is expressed as a ratio, called *heritability*, but it is not always clear what it means.

What does it mean to say that 80 percent of IQ is inherited? Technically, it means that a person's position in a group of people ranked by IQ would be unlikely to change by more than 20 percent if she had been raised differently. But that is not what we ordinarily mean by a "percentage of" something. The results of traditional twin studies, and similar types of studies, do not mean that your IQ is set by genetics and can only vary up or down by 20 percent after that. Nor does it mean that your IQ is set to a specific range by genetics and cannot rise or fall above or below that range due to environmental influences. The statistical measure derived from traditional studies only compares a person's IQ or other trait to the group of similar people also measured.

The techniques of modern genetics allow us to actually determine what genes different people have. This allows for two basic experimental approaches.

Two groups of people who exhibit different behaviors can have their DNA examined. Researchers look for individual genes, or small sets of genes, that differ between the groups. If all the members of one group exhibiting one type of behavior share a genetic pattern never found in the other group, then those genes are good candidates for causes of the behavior. In reality, this sort of perfect match rarely happens. Individual genes and even groups of genes most often influence behavior rather than completely controlling it. Having a specific gene may correlate with types of behavior, but it is unlikely to predict the behavior exactly.

The other method is to identify specific genes first, divide the subjects into groups based on different genes, and then compare their behavior. For example, the gene that produces a chemical that helps control the rate of serotonin reuptake has been identified. About half of the people studied have one version of the gene, about half have another. Subjects can be divided according to which version of the gene they have and then they can be tested for various personality traits. It has been shown that people with the different genes differ significantly on a number of personality measures.

Summary

The importance of the nervous system in generating behavior has long been recognized. However, the enormous complexity of the nervous system means that even as we learn tremendous amounts about how it works, we continue to be a long way away from understanding even the most basic facts about how it produces behavior.

Quiz

1. _____ is the historical idea that the nervous system produces behavior using complex versions of the reflex.

   (a) Reflexonomy

   (b) Reflexology

   (c) Cartesian psychology

   (d) None of the above

2. Which of the following is not an approach to understanding how psychological phenomena are produced by the nervous system?

   (a) Bottom-up approach

   (b) Modular approach

   (c) Cognitive neuroscience

   (d) Top-down approach

3. The _____ nervous system is made up of the brain and the spinal cord.

   (a) Central

   (b) Peripheral

   (c) Autonomic

   (d) Skeletal

4. Which structure of the central core of the brain is primarily involved with maintaining a sense of balance?

   (a) Thalamus

   (b) Cerebellum

   (c) Medulla

   (d) Pons

5. Which structure of the brain arouses the forebrain when information related to survival must be processed?

   (a) Pons

   (b) Thalamus

   (c) Reticular formation

   (d) Cerebellum

6. The _____ lobe of the cerebral cortex plays a major role in vision.

    (a) Occipital

    (b) Parietal

    (c) Temporal

    (d) Frontal

7. In the anatomy of a neuron, the _____ is the cellular nucleus, the _____ receives impulses from other cells, and the _____ carries impulses to other cells.

    (a) Dendrites, axon, soma

    (b) Soma, axon, dendrites

    (c) Axon, dendrites, soma

    (d) Soma, dendrites, axon

8. _____ occurs when a neurotransmitter causes the cell to pause activity; _____ occurs when a neurotransmitter activates an action potential in the receptor cell.

    (a) Excitation; inhibition

    (b) Inhibition; excitation

    (c) Both a & b

    (d) None of the above

9. Which of the following is not a method for monitoring the brain?

    (a) EEG

    (b) ERP

    (c) URP

    (d) fMRI

10. Which of the following is not a method used to determine how behavior is influenced by genetics?

    (a) Twin studies

    (b) Family studies

    (c) Adoption studies

    (d) All of the above are used.

PART THREE TEST

1. In signal detection theory, when a signal is reported as not being perceived and, in reality, is not present, it is classified as a _____.

   (a) Hit

   (b) Miss

   (c) False alarm

   (d) Correct rejection

2. Which theory of color vision explains color vision at the rod and cone level?

   (a) Retinex theory

   (b) Opponent-process theory

   (c) Trichromatic theory

   (d) All of the above

3. Which theory of color vision cannot explain color constancy?

   (a) Trichromatic theory

   (b) Retinex theory

   (c) Opponent-process theory

   (d) None of the above

4. In hearing, _____ theory is believed to be true for high-pitched sounds and _____ theory is true for low-pitched sounds.

   (a) Retinex, frequency

   (b) Vibration, pitch

   (c) Pitch, frequency

   (d) Frequency, pitch

5. The senses based on transducers in the skin, called the _____ senses, tell us about the relation between objects and our bodies.

   (a) Cutaneous

   (b) Vestibular

   (c) Somasthetic

   (d) Kinesthetic

6. Which part of the ear is filled with fluid and lined with cilia (tiny hairs) that move when membranes above and below them vibrate due to changes in fluid pressure?

   (a) Eardrum

   (b) Anvil

   (c) Hair cells

   (d) Cochlea

7. Our sense of flavor is really a combination of which senses?

   (a) Taste and smell

   (b) Smell and hearing

   (c) Vision and taste

   (d) Vision and smell

8. Which of the following is not a type of taste receptor?

   (a) Salt

   (b) Sweet

   (c) Bitter

   (d) None of the above

9. _____ depth cues are those that require both eyes in order to be perceived; _____ depth cues are those that require one eye in order to be perceived.

   (a) Binocular; monocular

   (b) Monocular; binocular

   (c) Size; linear

   (d) Linear; size

10. Which Gestalt principle states that when a figure is interrupted or incomplete, we tend to complete it?

    (a) Continuation

    (b) Proximity

    (c) Similarity

    (d) Closure

11. Which Gestalt principle states that stimuli that are close together are perceived as being grouped together?

    (a) Continuation

    (b) Proximity

    (c) Similarity

    (d) Closure

12. Which of the following is not a monocular depth cue?

    (a) Relative size

    (b) Closeness to the horizon

    (c) Convergence

    (d) Interposition

13. In _____ processing, in order to form a perception, we use our prior knowledge to organize and interpret sensations as soon as they are registered by the brain.

    (a) Bottom-up

    (b) Top-down

    (c) Depth

    (d) Lateral

14. Figure-ground separation occurs in which of the following perceptual systems?

    (a) Vision

    (b) Hearing

    (c) Both a and b

    (d) None of the above

15. _____ states that an object moving slowly across our visual field will be perceived to be more distant than an object moving quickly across our visual field.

    (a) Convergence

    (b) Linear perspective

    (c) Interposition

    (d) Motion parallax

16. Which factors play a large role in the top-down processing of sensory information?

    (a) Motivation

    (b) Perceptual set

    (c) Expectation

    (d) All of the above

17. The _____ lobe of the cerebral cortex plays a major role in higher reasoning and cognition.

    (a) Occipital

    (b) Parietal

    (c) Temporal

    (d) Frontal

18. In the anatomy of a neuron, the _____ carries impulses to other cells.

    (a) Dendrites

    (b) Axon

    (c) Soma

    (d) All of the above

19. The _____ is the master gland for the neuroendocrine system.

    (a) Pituitary gland

    (b) Pineal gland

    (c) Thyroid gland

    (d) Testes

20. The _____ is the large network of nerve cells that connects the two hemispheres of the brain.

    (a) Hypothalamus

    (b) Pineal gland

(c) Corpus callosum

(d) Temporal lobe

21. Too little of which neurotransmitter is linked to Parkinson's disease?

   (a) Serotonin

   (b) Dopamine

   (c) Norepinephrine

   (d) Endorphins

22. Too much of which neurotransmitter is linked to pain relief and runner's high?

   (a) Serotonin

   (b) Dopamine

   (c) Norepinephrine

   (d) Endorphins

23. Which brain structure is believed to be greatly involved in memory?

   (a) Hippocampus

   (b) Medulla

   (c) Reticular formation

   (d) Temporal lobe

24. Damage to structures in which area of the brain almost always lead to death?

   (a) Limbic system

   (b) Central core

   (c) Cerebral cortex

   (d) Corpus callosum

25. The advantage to the _____ is that it records highly specific information from a specific location in the brain.

   (a) Genetic analysis

   (b) EEGs

   (c) Individual ERP

   (d) fMRIs

FINAL EXAM

1. _____ science is science for the sake of understanding, and _____ science is science for the sake of problem solving.

   (a) Basic, basic

   (b) Basic, applied

   (c) Applied, basic

   (d) Applied, applied

2. Which of the following is a criterion of a good theory?

   (a) Fits the known facts

   (b) Makes predictions

   (c) Falsifiable

   (d) All of the above

3. A _____ is a testable prediction.

   (a) Hypothesis

   (b) Theory

   (c) Experimental group

   (d) Statistic

4. The operational definition for a dependent variable is how it is _____, and the operational definition for an independent variable is how it is _____.

   (a) Manipulated, measured

   (b) Measured, manipulated

   (c) Measured, confounded

   (d) Confounded, manipulated

5. A _____ sample is one in which the important subgroups of the population are present in the sample according to their percentages in the population.

   (a) Sub

   (b) Convenience

   (c) Random

   (d) Representative

6. _____ validity occurs when the measure yields comparable results to a different, already validated measure.

   (a) Face

   (b) Criterion

   (c) Content

   (d) Construct

7. To conduct a(n) _____ replication, a researcher will use different manipulations of the independent variable or different measures of the behavior of interest.

   (a) Exact

   (b) Confounding

   (c) Conceptual

   (d) All of the above

8. Manipulation of an independent variable and random assignment to groups are necessary for which type of research design?

   (a) True experiment

   (b) Correlational design

   (c) Case history

   (d) Survey

9. What method is used to reduce experimenter bias?

   (a) Careful examination of the laboratory

   (b) Blind study

   (c) A well-designed research design

   (d) All of the above

10. _____ variables are variables that the experimenter has not controlled for that still may affect the results.

    (a) Independent

    (b) Dependent

    (c) Confounding

    (d) Representative

11. The _____ technique is based on changing the cost of a behavior once you have already agreed to comply.

    (a) Low-ball

    (b) Door-in-the-face

    (c) Foot-in-the-door

    (d) That's-not-all

12. According to Milgram's studies of obedience, the farther you are from the victim the _____ likely you are to obey an order to harm them.

    (a) More

    (b) Less

    (c) Equally

    (d) All of the above

13. Which stage of a relationship is considered the "settling down" stage?

    (a) Initial attraction

    (b) Building

    (c) Consolidation

    (d) All of the above

14. If the barriers to leaving a relationship are low, the likelihood of leaving _____.

    (a) Decreases

    (b) Increases

    (c) Stays the same

    (d) Is negotiated

15. The _____ technique induces compliance by adding a benefit or dropping the cost while the customer considers the deal.

    (a) Low-ball

    (b) Door-in-the-face

(c) Foot-in-the-door

(d) That's-not-all

16. People who are high in their need for cognition tend to be _____ to persuade than those who are low in their need for cognition.

(a) Harder

(b) Easier

(c) About the same

(d) Identical

17. Piaget argued that the cognitive task of the _____ is for a child to expand her world beyond the limits of her immediate perceptions.

(a) Sensorimotor Stage

(b) Formal operational stage

(c) Pre-operational stage

(d) Concrete operational stage

18. When the caregiver is not consistent in appropriately responding to the infant's needs, the infant can develop a(n) _____ attachment to the caregiver.

(a) Secure

(b) Resistant

(c) Avoidant

(d) Disorganized

19. Erikson's challenge for middle adulthood revolves around developing _____.

(a) Generativity vs. stagnation

(b) Autonomy vs. shame

(c) Identity vs. role confusion

(d) Intimacy vs. isolation

20. If an identity crisis is absent and personal commitment is absent, which identity orientation is likely to result?

(a) Diffusion

(b) Moratorium

(c) Achievement

(d) Foreclosure

21. If an identity crisis is absent and personal commitment is present, which identity orientation is likely to result?

    (a) Diffusion

    (b) Moratorium

    (c) Achievement

    (d) Foreclosure

22. According to Erikson, the_____ stage is marked by the opportunity to develop an integrated sense of self as distinct from other people.

    (a) Generativity vs. stagnation

    (b) Autonomy vs. shame

    (c) Identity vs. role confusion

    (d) Intimacy vs. isolation

23. _____ validity involves determining the test's ability to cover the complete range of material that it is supposed to cover.

    (a) Predictive

    (b) Content

    (c) Test-retest

    (d) Criterion

24. How many IQ scores can be derived from the WAIS?

    (a) Zero

    (b) One

    (c) Two

    (d) Three

25. Studies have shown that people with higher IQs tend to process information _____ than do people with lower IQs.

    (a) Slower

    (b) Faster

    (c) At the same speed

    (d) None of the above

26. According to Sternberg's Triarchic theory of intelligence, which of the following is not a component of analytic intelligence?

    (a) Creative component

    (b) Metacomponents

(c) Performance components

(d) Knowledge-acquisition components

27. According to Gardner's theory of multiple intelligences, the ability to understand one's self is _____ intelligence.

(a) Naturalistic

(b) Interpersonal

(c) Intrapersonal

(d) Spatial

28. According to Gardner's theory of multiple intelligences, the ability to manipulate abstract symbols is _____ intelligence.

(a) Naturalistic

(b) Mathematical

(c) Intrapersonal

(d) Spatial

29. Those categorized with _____ retardation typically fall within the 50 to 70 IQ range, are generally able to complete sixth-grade academic work, and may hold a job in a supportive setting.

(a) Mild

(b) Moderate

(c) Severe

(d) All of the above

30. Which defense mechanism did Freud consider to be the ego's first line of defense?

(a) Identification

(b) Denial

(c) Regression

(d) Repression

31. Freud argued that the failure to resolve the conflicts at any stage of psychosexual personality development can result in _____.

(a) Fixation

(b) Regression

(c) Fantasy

(d) Intellectualization

32. According to Jung, personality is made up of both a personal unconscious and a(n) _____ unconscious.

    (a) Rational

    (b) Collective

    (c) Superego

    (d) Emotive

33. According to Allport, _____ traits are those traits that represent the major characteristics of a person.

    (a) Central

    (b) Cardinal

    (c) Primary

    (d) Secondary

34. Which of the following is not considered one of the "big five" personality traits?

    (a) Openness

    (b) Extraversion

    (c) Neuroticism

    (d) Altruism

35. According to Eysenck's model of personality structure, _____ are made up of correlated habitual responses.

    (a) Traits

    (b) Types

    (c) Single responses

    (d) Habits

36. What familial factors influence our tendency to engage in unhealthy behaviors?

    (a) Genetics

    (b) Learning

    (c) Both a and b

    (d) None of the above

37. Which of the following is a component of our stress reaction?

    (a) Physiological

    (b) Emotional

(c) Behavioral

(d) All of the above

38. _____ coping strategies attempt to cope with stress by focusing on changing the thoughts about the stressful situation.

(a) Emotion-focused

(b) Problem-focused

(c) Cognitive-focused

(d) Biologically focused

39. _____ coping strategies attempt to cope with stress by altering the unpleasant emotions caused by the stressful situation.

(a) Emotion-focused

(b) Problem-focused

(c) Cognitive-focused

(d) Biologically focused

40. According to the Protection Motivation Theory, which of the following is not a factor that leads us to continue to engage in unhealthy behaviors?

(a) Decreased perceived susceptibility to the health threat

(b) Decreased perceived severity of the health threat

(c) Decreased self-efficacy of behavior change

(d) Barriers to behavior change outweigh the benefits.

41. Unhealthy behaviors are typically practiced first during _____.

(a) Infancy

(b) Childhood

(c) Adolescence

(d) Adulthood

42. The behavioral response to stress is termed _____.

(a) Coping

(b) Fleeing

(c) Susceptibility

(d) Severity

43. _____ occurs when a person's personality fractures into two or more distinct identities.

    (a) Bipolar disorder

    (b) Obsessive-compulsive disorder

    (c) Schizophrenia

    (d) Dissociative identity disorder

44. Clinical researchers and practitioners using the _____ are focused on biological causes and pharmacological treatments.

    (a) Medical model

    (b) Biopsychosocial

    (c) Clinical

    (d) None of the above

45. The DSM is a _____ system for psychological disorders.

    (a) Drug

    (b) Medical

    (c) Treatment

    (d) Classification

46. Which of the following is not classified by the DSM as a psychological disorder typically diagnosed in infancy, childhood, or adolescence?

    (a) Attention-deficit disorder

    (b) Schizophrenia

    (c) Autism

    (d) Learning disabilities

47. Panic disorder, obsessive-compulsive disorder, and phobias are considered to be _____ disorders.

    (a) Somatoform

    (b) Mood

    (c) Anxiety

    (d) Personality

48. Delusions of _____ involve the belief that you are especially important or that you are a famous person.

    (a) Reference

    (b) Grandeur

(c) Persecution

    (d) Control

49. Which psychological disorder first attracted Freud's attention?

    (a) Anxiety disorder

    (b) Depression

    (c) Schizophrenia

    (d) Somatization disorder

50. _____ is the technique in which the therapist encourages the patient to say whatever is on his mind without self-censoring.

    (a) Free association

    (b) Reactance

    (c) Transference

    (d) Exposure

51. Which type of therapy uses techniques based on unconditional positive regard?

    (a) Psychoanalysis

    (b) Client-centered therapy

    (c) Behavioral therapy

    (d) Systematic desensitization

52. _____ techniques are often used during skills training therapies.

    (a) Insight therapy

    (b) Operant conditioning

    (c) Classical conditioning

    (d) Observational learning

53. Which form of cognitive therapy assumes that irrational thoughts stem from cognitive distortions of reality?

    (a) Rational-emotive therapy

    (b) Beck's cognitive therapy

    (c) Freud's psychoanalysis

    (d) Cognitive-behavioral therapy

54. Mood stabilizers target which neurotransmitters to ease the symptoms of depression?

    (a) Endorphins

    (b) Dopamine

(c) Norepinephrine

(d) GABA

55. What factors account for the rise in group therapies after World War II?

(a) Increased demand for therapy

(b) Shortage of therapists

(c) Both a and b

(d) None of the above

56. The presence of _____ things spurs us to specific actions, while the presence of _____ things does not.

(a) Unpleasant, pleasant

(b) Pleasant, unpleasant

(c) Unpleasant, unpleasant

(d) Pleasant, pleasant

57. A classification scheme that takes the difference between sentiments and moods seriously leads us to a consideration of the _____ of emotions.

(a) Classification

(b) Structure

(c) Function

(d) All of the above

58. The _____ perspective suggests that the basic emotions are those with survival value.

(a) Behavioral

(b) Cognitive

(c) James-Lange

(d) Evolutionary

59. According to the evolutionary perspective of emotion, which emotions do not have potential survival value?

(a) Jealousy

(b) Anger

(c) Fear

(d) None of the above

60. If someone were trying to deceive you, which type of emotional expression would give you the best indicators of the deception?

   (a) Facial expressions

   (b) Body language

   (c) Verbal communication

   (d) Gestures

61. It is easier to determine the emotional state of someone who _____ your cultural heritage.

   (a) Has heard of

   (b) Differs with

   (c) Shares

   (d) Appreciates

62. The Freudian perspective and the behavioral perspective differ on which aspect(s) of the study of motivation?

   (a) The importance of motivation

   (b) How to research the influence of motivation

   (c) The definition of motivation

   (d) All of the above

63. The _____ perspective was one of the earliest psychological perspectives to study motivation.

   (a) Psychoanalytic

   (b) Behavioral

   (c) Cognitive

   (d) Humanistic

64. Operant conditioning theory states that motivation that influences the environmental contingency of an event is_____.

   (a) Rewarding

   (b) Punishing

   (c) Neutral

   (d) Any of the above

65. _____ drives are those that satisfy a biological need; _____ drives are those that do not satisfy a biological need directly.

    (a) Primary; primary

    (b) Primary; secondary

    (c) Secondary; primary

    (d) Secondary; secondary

66. Which of the following does not influence the improvement of goal-directed behavior?

    (a) Commitment

    (b) Experience discrepancy

    (c) Payment

    (d) Feedback

67. Food preferences are largely determined by _____.

    (a) Cultural preferences

    (b) Genetics

    (c) Habit

    (d) What is on the menu

68. Ethologists argue that the win-stay strategy works well for _____.

    (a) Foragers

    (b) Prey

    (c) Predators

    (d) All of the above

69. _____ conditioning conditions new stimuli to old responses, while _____ conditioning conditions new responses to old stimuli.

    (a) Classical, operant

    (b) Operant, classical

    (c) Classical, classical

    (d) Operant, operant

70. Behavior conditioned to a red light that also appears in the presence of an orange light is an example of which conditioning phenomena?

    (a) Extinction

    (b) Stimulus discrimination

(c) Stimulus generalization

(d) Spontaneous recovery

71. Behavior conditioned to a red light that is absent in the presence of an orange light when the orange light was presented alternatively during conditioning is an example of which conditioning phenomena?

(a) Extinction

(b) Stimulus discrimination

(c) Stimulus generalization

(d) Spontaneous recovery

72. Response generalization is found in _____ conditioning, but not in _____ conditioning.

(a) All, modeling

(b) Social learning, operant

(c) Classical, operant

(d) Operant, classical

73. Genetically preprogrammed behaviors are also known as _____.

(a) Fixed action patterns

(b) Fixed action potentials

(c) Fixed responses

(d) Functional fixedness

74. The theories of language acquisition are deeply rooted in the _____ question.

(a) Nature-nurture

(b) Genetics

(c) Mind-body

(d) None of the above

75. Chomsky argued that the brain has a structure that analyzes automatically the components of the speech a child hears. This structure is known as the _____.

(a) Limbic system

(b) Occipital lobe

(c) Hippocampus

(d) Language acquisition device

76. The fact that there is no built-in relationship between a word and the concept it represents indicates that language is _____.

    (a) Generative

    (b) Symbolic

    (c) Semantic

    (d) Structured

77. A _____ vocabulary contains the words we generate.

    (a) Receptive

    (b) Symbolic

    (c) Productive

    (d) Semantic

78. A _____ vocabulary contains the words we understand.

    (a) Receptive

    (b) Productive

    (c) Symbolic

    (d) Generative

79. Which of the following is not true about the use of gestures?

    (a) Gestures can accompany verbal communication.

    (b) Gestures can be used without verbal communication.

    (c) Gestures may have a genetic component.

    (d) Gestures rarely accompany verbal communication.

80. _____ rehearsal is simply repeating something to ourselves without giving it much additional thought.

    (a) Semantic

    (b) Maintenance

    (c) Elaborative

    (d) Phonetic

81. _____ is the process we use to distinguish memories of real events versus imagined events.

    (a) Amnesia

    (b) Forgetting

(c) Reality monitoring

(d) Repressed memories

82. _____ are simple techniques that can greatly improve anyone's powers of recall.

(a) Heuristics

(b) Mnemonic devices

(c) Automatic encoding

(d) None of the above

83. _____ memory involves using external cues, such as pictures on the wall or your location in the room, to aid in the retrieval of information.

(a) Mnemonic devices

(b) Visual imagery

(c) State-dependent memory

(d) Context-dependent memory

84. _____ is the loss of memories over a specific time span, typically resulting from brain damage caused by accident, stroke, or disease.

(a) Amnesia

(b) Repressed memory

(c) Reality monitoring

(d) State-dependent memory

85. Which of the following memory store type has a separate storage system for each of the five senses?

(a) Sensory memory

(b) Short-term memory

(c) Long-term memory

(d) Flashbulb memory

86. A(n) _____ is a short-cut used in problem solving that saves time, but does not always lead to a correct solution.

(a) Heuristic

(b) Algorithm

(c) Trial-and-error strategy

(d) Rule

87. _____ inferences are needed in decision-making, in situations where we do not always know things for certain.

    (a) Abductive

    (b) Deductive

    (c) Inductive

    (d) All of the above

88. The _____ heuristic is where people overvalue the similarity of an individual event to other events, ignoring the overall likelihood of that type of event.

    (a) Availability

    (b) Uncertainty

    (c) Risk

    (d) Representativeness

89. A _____ problem-solving strategy involves beginning at the endpoint and working your way to the beginning.

    (a) Backward search

    (b) Subgoal analysis

    (c) Mental set

    (d) Means-end analysis

90. A(n) _____ inference is the strongest form of inference.

    (a) Deductive

    (b) Abductive

    (c) Inductive

    (d) Heuristic

91. The absolute threshold is the intensity level where the subject detects the stimulus _____ of the time.

    (a) 25 percent

    (b) 50 percent

    (c) 80 percent

    (d) 95 percent

92. In signal detection theory, when a signal is reported as being perceived and, in reality, is present, it is classified as a _____.

    (a) Hit

    (b) Miss

(c) False alarm

(d) Correct rejection

93. The _____ is the part of the eye responsible for most of its power to focus light into the eye.

   (a) Pupil

   (b) Lens

   (c) Cornea

   (d) Fovea

94. The _____ is where the axons from retinal nerve cells in each eye are gathered into a single large cord, which carries impulses from the retina into the brain.

   (a) Lens

   (b) Fovea

   (c) Optic nerve

   (d) Cones

95. Being able to hear and understand a waiter's description of the specials in a noisy restaurant is an example of _____.

   (a) Figure-ground separation

   (b) Proximity

   (c) Relative size

   (d) None of the above

96. Which of the following is not a monocular depth cue?

   (a) Retinal disparity

   (b) Closeness to the horizon

   (c) Convergence

   (d) All of the above

97. Pictorial depth cues are _____ depth cues.

   (a) Monocular

   (b) Binocular

   (c) Both a and b

   (d) None of the above

98. The _____ nervous system contains the nerves that communicate with and control the body systems that manage basic life functions such as respiration, circulation, reproduction, and digestion.

    (a) Skeletal

    (b) Motoric

    (c) Sensory

    (d) Autonomic

99. Which of the following is true of the human nervous system?

    (a) Glia cells exist only in the brain.

    (b) All neurons communicate via synapses.

    (c) All neurons transmit action potentials.

    (d) None of the above.

100. Genetic studies have shown that a gene influencing serotonin affects _____.

    (a) Learning

    (b) Memory

    (c) Personality

    (d) None of the above

ANSWERS

Chapter 1

1. D
2. B
3. A
4. C
5. D
6. C
7. D
8. C
9. A
10. B

Chapter 2

1. B
2. C
3. D
4. A
5. C
6. B
7. D
8. B

9. A
10. B

Chapter 3

1. B
2. D
3. A
4. B
5. C
6. D
7. B
8. A
9. D
10. B

Chapter 4

1. B
2. C
3. C
4. A
5. D

6. D
7. A
8. D
9. C
10. B

Chapter 5

1. D
2. C
3. A
4. C
5. B
6. A
7. D
8. C
9. B
10. B

Chapter 6

1. D
2. A

3. C
4. C
5. A
6. D
7. B
8. D
9. D
10. B

Chapter 7

1. D
2. B
3. B
4. A
5. A
6. C
7. B
8. C
9. D
10. C

Chapter 8

1. B
2. D
3. A
4. B
5. C
6. D
7. A
8. C
9. D
10. A

Part One Test

1. A
2. B
3. D
4. C
5. C
6. A
7. C
8. B
9. D
10. B
11. C
12. B
13. C
14. A
15. C
16. D
17. A
18. D
19. A
20. B
21. B
22. D
23. B
24. D
25. C

Chapter 9

1. E
2. A
3. D

4. B
5. C
6. A
7. D
8. B
9. A
10. C

Chapter 10

1. D
2. B
3. D
4. C
5. A
6. D
7. C
8. B
9. B
10. D

Chapter 11

1. D
2. C
3. A
4. B
5. A
6. D
7. C
8. D
9. B
10. C

Chapter 12

1. D
2. A
3. C
4. D
5. A
6. B
7. C
8. D
9. B
10. A

Chapter 13

1. B
2. C
3. D
4. B
5. C
6. A
7. D
8. A
9. D
10. A

Chapter 14

1. C
2. D
3. B
4. A
5. C
6. B

7. D
8. B
9. D
10. A

Part Two Test

1. D
2. B
3. D
4. B
5. B
6. A
7. C
8. A
9. D
10. C
11. A
12. B
13. A
14. C
15. D
16. B
17. B
18. C
19. A
20. C
21. C
22. A
23. B
24. D
25. B

Chapter 15

1. A
2. C
3. B
4. C
5. D
6. A
7. B
8. D
9. C
10. D

Chapter 16

1. B
2. D
3. B
4. C
5. A
6. B
7. D
8. C
9. A
10. D

Chapter 17

1. B
2. D
3. A
4. B
5. C
6. A

7. D
8. B
9. C
10. D

Part Three Test

1. D
2. C
3. C
4. C
5. A
6. D
7. A
8. D
9. A
10. D
11. B
12. C
13. B
14. C
15. D
16. D
17. D
18. B
19. A
20. C
21. B
22. D
23. A
24. B
25. C

Final Exam

1. B
2. D
3. A
4. B
5. D
6. B
7. C
8. A
9. B
10. C
11. A
12. A
13. C
14. B
15. D
16. A
17. C
18. C
19. A
20. A
21. B
22. C
23. B
24. D
25. B
26. A
27. C
28. B
29. A
30. D
31. A

32. B
33. A
34. D
35. A
36. C
37. D
38. C
39. A
40. C
41. C
42. A
43. D
44. A
45. D
46. B
47. C
48. B
49. D
50. A
51. B
52. D
53. B
54. C
55. C
56. A
57. C
58. D
59. D
60. B
61. C
62. B
63. A

64. D	77. C	90. A
65. B	78. A	91. B
66. C	79. D	92. A
67. A	80. B	93. C
68. C	81. C	94. C
69. A	82. B	95. A
70. C	83. D	96. B
71. B	84. A	97. A
72. D	85. A	98. D
73. A	86. A	99. A
74. A	87. C	100. C
75. D	88. D	
76. B	89. A	

APPENDIX

Resources for Learning

As you probably have guessed, there is tons more psychology out there in addition to what we put into *Psychology Demystified*. Much of this information can be discovered through further psychology courses. There are many excellent books as well, written for a lay audience, about various topics in psychology. We have listed a small number in this appendix as our recommendations for further reading. We had three criteria: They had to be well-written, not require a background in psychology, and take an interesting position on a psychological topic.

The second author of this book has a web site that has, among other things, links to various other sites discussing psychology. Some of these links are about more technical topics, but there are sites of general interest as well. Go to *www.unc.edu/~skemp/Science/smkSciForum.html#links* for more information.

We want to keep learning as well. We'd love to hear back from you with any suggestions you have for improving the book or the web site. Send comments or questions to *steve_kemp@unc.edu*.

Recommended Reading

The following books are a small selection of readings for those interested in finding out more about various specific topics in psychology. Carol Tavris' (2000) *Psychobabble and Biobunk* is a terrific book about how to be an intelligent consumer of mass media reports about social science research. Darrell Huff's (1954) *How to Lie with Statistics* makes statistics both fun and interesting. Freud (1989) and Skinner (1974) are both the best sources for their own views, and both books are in paperback. For those interested in the relationship between psychology, health, stress, and coping, Reynolds Price's (1982) autobiographical book, *A Whole New Life: An Illness and a Healing*, is fascinating. Steven Pinker's (2002) *The Blank Slate: The Modern Denial of Human Nature* can be pretty technical, but he has a fascinating view of language. All the other books listed next are either books the authors have read and recommend, or books recommended to us by folks we trust.

REFERENCES

Cialdini, R. B. (1984). *Influence: The Psychology of Persuasion.* New York: William Morrow and Company, Inc.

Dawkins, Richard. (1989). *The Selfish Gene.* Oxford: Oxford University Press.

De Waal, F. (2005). *Our Inner Ape : A Leading Primatologist Explains Why We Are Who We Are.* New York: Riverhead.

Freud, S. (1989). *Introductory Lectures on Psycho-Analysis.* (James Strachey, Trans.). New York: Liveright Publishing Corporation.

Gould, S. J. (1996). *The Mismeasure of Man.* New York: W.W. Norton & Company.

Huff, D. & Geis, I. (1954). *How to Lie with Statistics.* (Norton, ppbk., 1993). New York: W.W. Norton & Company.

Miller, G. (2000). *The Mating Mind : How Sexual Choice Shaped the Evolution of Human Nature.* (Anchor, ppbk., 2001). New York: Doubleday.

Norman, D. A. (1988). *The Psychology of Everyday Things.* New York: Basic Books.

Pinker, S. (2002). *The Blank Slate : The Modern Denial of Human Nature.* (Penguin, ppbk., 2003). New York: Viking.

Price, R. (1982). *A whole new life: An illness and a healing.* New York, NY: Penquin/Plume.

Skinner, B. F. (1974). *About Behaviorism.* New York: Random House, Inc.

Szasz, T. S. (1974). *The Myth of Mental Illness: Foundations of a Theory of Personal Conduct.* (Revised Edition.) New York: Harper & Row, Publishers, Inc.

Tavris, C. (2000). *Psychobabble and Biobunk: Using Psychology to Think Critically About Issues in the News.* (2nd Edition). Upper Saddle River, NJ: Prentice Hall.

GLOSSARY

Psychology is filled with specialized terms. Here are terms we have used, and defined, in *Psychology Demystified.*

Quick Quotes: Psychology

"The science that explains what everyone knows in language that no one can understand." Bode, B. H. (1922). "What Is Psychology?" *The Psychological Review,* 29, 250–258. (For the record, Professor Bode does not claim to have invented this definition. It's just something he heard.)"

A

Applied science   Science for the sake of problem solving. (*Chapter 1, methods*)

Associationism   A philosophical approach that assumes behavior can be understood in terms of environmental circumstances causing actions. (*Chapter 11, learning*)

Attachment   Enduring emotional ties that bond a child to his or her primary caregiver. (*Chapters 3 and 10, development, motivation*)

Attitude   The association between one's thoughts and an evaluation. (*Chapter 2, social*)

Attribution   The process of assigning causes to behavior. (*Chapter 2, social*)

B

Basic science   Science for the sake of understanding. (*Chapter 1, methods*)

Biases and heuristics   Systematic errors of reasoning, judgment, and decision-making. (*Chapters 2 and 14, social, thinking*)

Blinded study   A study in which either the researcher or the subject or both do not know to which group the subject has been assigned. (*Chapter 1, methods*)

C

Central nervous system   The brain and the spinal cord. (*Chapter 17, neural*)

Classical conditioning   A method for producing associative learning based on the work of Ivan Pavlov. (*Chapter 11, learning*)

Cognitive dissonance   An inconsistency between thoughts or behaviors that leads to tension, and the changing of thoughts or behaviors to ease the tension. (*Chapter 2, social*)

Cognitivism   A theoretical approach that assumes the mind works like a computer. (*Chapter 14, thinking*)

Conditioning   Using specialized methods (other than teaching) to produce learning. (*Chapter 11, learning*)

Control/comparison group   The group that does not receive the manipulation of the independent variable. (*Chapter 1, methods*)

Correlation   A measure of the relation between two sets of values. The result of the statistic is the correlation coefficient. (*Chapter 1, methods*)

Cutaneous senses   The senses of touch, temperature, and pain. (*Chapter 15, sensation*)

D

Data   The numbers (or other symbols) produced by observation or measurement in a scientific study. (*Chapter 1, methods*)

Defense mechanisms   Mental strategies used to defend the ego in the daily conflict between id impulses and the superego's demands to deny them. (*Chapter 5, personality*)

Dependent measure/variable   The research variable that is measured in order to determine the results of interest. (*Chapter 1, methods*)

Depression   A mood disorder in which a person experiences severe sadness and hopelessness. (*Chapter 7, disorders*)

Descriptive statistics   Summary information about the data, such as sums and averages, that provide one or two numbers to tell us about a larger group of numbers. (*Chapter 1, methods*)

Diagnostic and Statistical Manual of Mental Disorders (DSM)   The "bible" of the psychiatric profession, listing all officially recognized psychiatric problems. (*Chapter 7, disorders*)

E

Ego   The reality-based aspect of the self that represents a person's view of physical and social reality, or his or her conscious beliefs about the causes and consequences of behavior. (*Chapter 5, personality*)

Encoding   Transforming a piece of information into a form storable in memory. (*Chapter 13, memory*)

Ethology   An approach to studying behavior that focuses on species differences and evolution. (*Chapter 11, learning*)

Experimental group   The group that is exposed to the presence, or manipulation, of the independent (causal) variable. (*Chapter 1, methods*)

F

Family studies   Studies that examine blood relatives to determine how much they resemble each other on a specific trait. (*Chapter 17, neural*)

Feature   Aspects of an object used to categorize the object. (*Chapter 14, thinking*)

Fight/flight   A class containing two different mammalian reactions to danger. (*Chapter 9, emotion*)

Figure-ground   The distinction between the object we are viewing and the background behind the figure. (*Chapter 16, perception*)

Fundamental attribution error   The tendency to attribute behavior to internal rather than external causes. (*Chapter 2, social*)

G

Gestalt   Holistic approach to perception based on the notion that "the whole is greater than the sum of its parts." (*Chapter 16, perception*)

Gray matter   The portions of the brain that consist of a large proportion of cell bodies and dendrites, associated with behavior and thinking. (*Chapter 17, neural*)

H

Hypothesis   A tentative statement about what happens in a given situation; a testable prediction, derived from theory. (*Chapter 1, methods*)

I

Id   The primitive, unconscious part of the personality; the storehouse of fundamental drives. (*Chapter 5, personality*)

Independent variable   The variable in an experiment that is manipulated or used to define comparison groups; the putative cause. (*Chapter 1, methods*)

Inferential statistics   Statistics that allow the researcher to draw conclusions about populations from samples used in a study. (*Chapter 1, methods*)

Interpersonal attraction   Positive feelings toward another, tending toward partner selection and/or affiliation. (*Chapter 2, social*)

Introspection   A technique in which subjects are asked to look within themselves and report their observations. (*Chapter 14, thinking*)

J

Just noticeable difference (jnd)   Also known as the *difference threshold*. The smallest detectible change in sensory energy. (*Chapter 15, sensation*)

L

Language acquisition device   A theorized structure in the brain that automatically analyzes the components of the speech a child hears. (*Chapter 12, language*)

Latent learning   When organisms learn a new behavior but do not perform that behavior until they have an incentive. (*Chapter 11, learning*)

M

Mental retardation   When the normal cognitive and intellectual development of the child slows, ceases, or reverses. (*Chapter 7, disorders*)

Mnemonic devices   Simple techniques that can greatly improve a person's powers of recall. (*Chapter 13, memory*)

Modeling   The ability to learn behaviors by observing others' behavior. (*Chapter 11, learning*)

Mood   Short-term emotional states not obviously focused on any one thing. (*Chapter 9, emotion*)

Motivational biases   Occurs when something motivates us internally to make a particular attribution. (*Chapter 2, social*)

Multiple personality disorder   Now called *dissociative identity disorder,* where one set of personal identity characteristics is temporarily forgotten and another set (or sets) is adopted. (*Chapter 7, disorders*)

N

Nerve   Bundles of neurons that communicate from one part of the body to another. (*Chapter 17, neural*)

Neurons Specialized cells that conduct signals from one place to another. (*Chapter 17, neural*)

O

Operant conditioning A method for producing learning based on the consequences of one's behavior. (*Chapter 11, learning*)

Operational definition Specifies the procedures used to produce or measure a variable. (*Chapter 1, methods*)

P

Parsimony With all else being equal, the better theory is the theory that explains reality in the simplest terms using the fewest number of assumptions. (*Chapter 1, methods*)

Perception How we turn raw sensation into useful information about the world. (*Chapter 16, perception*)

Personality disorders Mild psychiatric problems where normal personality trends are abnormally pervasive, persistent, and resist treatment. (*Chapter 7, disorders*)

Persuasion The study of how attitudes and opinions are changed. (*Chapter 2, social*)

Phobias Intense irrational fears of specific objects, activities, or situations that interfere with daily life. (*Chapter 7, disorders*)

Prototype An example that best represents a category. (*Chapter 14, thinking*)

Punishment A consequence that causes a behavior to be suppressed or decrease in frequency. (*Chapter 11, learning*)

Q

Quasi-experiment A study with all the features of an experiment except that subjects are not randomly assigned to groups; often used for field studies. (*Chapter 1, methods*)

R

Random sample Sample where every person in the population of interest has an equal chance of being included. (*Chapter 1, methods*)

Recall Retrieving specific pieces of information from memory, often guided by retrieval cues. (*Chapter 13, memory*)

Recognition Deciding whether or not you have ever encountered a particular stimulus. (*Chapter 13, memory*)

Reinforcement A consequence that changes the frequency of a behavior. Positive reinforcement results in an increase in frequency. (*Chapter 11, learning*)

Reinforcement schedule A set of rules for providing reinforcement depending on specific patterns of behavior; used to maintain, increase, or decrease the rate of responding. (*Chapter 11, learning*)

Reliability A variable is reliable if it consistently yields the same results every time it is measured. (*Chapter 1, methods*)

Repeatability The ability of any experiment to be repeated. (*Chapter 1, methods*)

Replicability The ability of any experiment to be repeated with the same results. (*Chapter 1, methods*)

Representation A hypothetical brain state that acts like a symbol, standing in for something in the real world. (*Chapter 12, language*)

Representative sample A sample in which the important subgroups of the population are represented in the sample according to their percentages in the population. (*Chapter 1, methods*)

Repression A Freudian mechanism whereby unconscious desires are kept from conscious awareness. (*Chapter 5, personality*)

S

Schemas, scripts, and frames Theoretical mechanisms that provide organization and structure to long-term memory, aiding in retrieval and thinking. (*Chapters 13 and 14, memory, thinking*)

Shaping   A technique for training novel behaviors using gradual approximations to the desired behavior. (*Chapter 11, learning*)

Standardization   The development of uniform procedures for giving and scoring a test. (*Chapter 4, intelligence*)

Superego   The storehouse of an individual's values, derived from society and parents, etc. (*Chapter 5, personality*)

Synapse   The gap between neurons that neural signals must cross, commonly using chemicals called neurotransmitters. (*Chapter 17, neural*)

T

Talk therapy   Any type of psychotherapy that relies upon talk between the patient and therapist for its therapeutic effect. (*Chapter 8, therapies*)

Theory of Multiple Intelligences   Posits that there are eight types of intelligence that involve a collection of abilities working together. (*Chapter 4, intelligence*)

Twin studies   A way to examine genetic influences by studying identical and fraternal twins. (*Chapters 4 and 17, intelligence, neural*)

V

Validity   A variable is valid if it measures what it is supposed to measure. (*Chapter 1, methods*)

W

White matter   The portions of the brain that consist of a large proportion of axons, associated with communication between different parts of the brain. (*Chapter 17, neural*)

INDEX